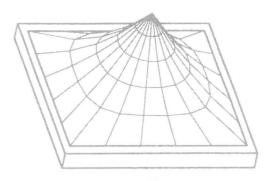

The Tensioned Fabric Roof

The Tensioned Fabric Roof

Craig G. Huntington

Library of Congress Cataloging-in-Publication Data

Huntington, Craig G.
    The tensioned fabric roof / Craig G. Huntington.
      p. cm.
    Includes bibliographical references and index.
    ISBN 0-7844-0428-3
    1. Roofs, Fabric—Design and construction. I. Title.

TH2449.H86   2003
695—dc22   2003068819

Unless otherwise noted, all art is used with permission of the author.

Published by American Society of Civil Engineers
1801 Alexander Bell Drive
Reston, Virginia 20191
www.asce pubs.asce.org

Any statements expressed in these materials are those of the individual authors and do not necessarily represent the views of ASCE, which takes no responsibility for any statement made herein. No reference made in this publication to any specific method, product, process, or service constitutes or implies an endorsement, recommendation, or warranty thereof by ASCE. The materials are for general information only and do not represent a standard of ASCE, nor are they intended as a reference in purchase specifications, contracts, regulations, statutes, or any other legal document.

ASCE makes no representation or warranty of any kind, whether express or implied, concerning the accuracy, completeness, suitability, or utility of any information, apparatus, product, or process discussed in this publication, and assumes no liability therefore. This information should not be used without first securing competent advice with respect to its suitability for any general or specific application. Anyone utilizing this information assumes all liability arising from such use, including but not limited to infringement of any patent or patents.

ASCE and American Society of Civil Engineers—Registered in U.S. Patent and Trademark Office.

Photocopies: Authorization to photocopy material for internal or personal use under circumstances not falling within the fair use provisions of the Copyright Act is granted by ASCE to libraries and other users registered with the Copyright Clearance Center (CCC) Transactional Reporting Service, provided that the base fee of $18.00 per article is paid directly to CCC, 222 Rosewood Drive, Danvers, MA 01923. The identification for ASCE Books is 0-7844-0428-3/04/ $18.00. Requests for special permission or bulk copying should be addressed to Permissions & Copyright Dept., ASCE.

Cover photo: An evening at the Weber Point Events Center; Stockton, CA. Photograph by Don Douglas; used with permission.

Copyright © 2004 by the American Society of Civil Engineers.
All Rights Reserved.
Library of Congress Catalog Card No: 2003068819
ISBN 0-7844-0428-3
Manufactured in the United States of America.

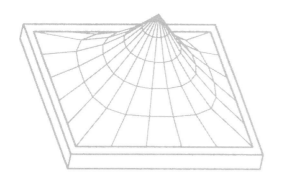

Contents

Foreword		vii
Preface		ix
Acknowledgments		xiii
1	The Tensioned Fabric Roof	1
2	Elements of Form and Design	11
3	Tensioned Fabric Structural Systems	21
4	Materials	57
5	Form Finding and Analysis	85
6	Connections and Detailing	101
7	Fabrication and Erection	133
8	Nonstructural Performance Parameters	155
9	The Contemporary Fabric Structures Industry	173
References		185
Project Credits		191
Index		197

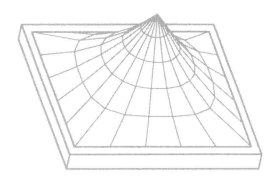

Foreword

Craig Huntington is one of a handful of structural engineers in the United States qualified and experienced in the analysis and design of tensile membrane structures. These unique lightweight and free-flowing canopies represent, more than most other structures, the merger of architectural form and engineering principles. In writing this remarkable book, Mr. Huntington is fortunate to be able to call upon an undergraduate degree in architecture as well as a Master of Engineering from the University of California at Berkeley.

Except in really long-span structures, the structural system of a traditional commercial or institutional building may or may not be reflected in its form. It usually *is*, for reasons of logic and simplicity, but it is by no means an inherent requirement. In tensioned fabric roofs, the subject of this book, the membrane structure *is* the form!

As the author points out, these forms are almost always visually very attractive (some say seductive), and I feel certain that his education in architecture helped bring him to a career in tensile membrane design.

The Tensioned Fabric Roof is a wonderfully unique and complete work that fills a wide gap in the currently available literature on membrane structures. Prior to its publication, there were (and still are) basically three types of reference materials available to those interested in the subject matter. First, there are many promotional brochures and booklets being produced annually by material suppliers and fabricators. These show the physical attributes and properties of materials and provide many pictures of a great variety of finished works. Second, there are a few highly technical engineering textbooks that examine the complicated mathematics underlying the nonlinear behavior of large displacement flexible membranes. (These texts are so esoteric that even most practicing membrane specialists do not fully comprehend them!) Third, there is a considerable number of what I refer to as "coffee-table" books. These expensive folio format books have beautiful pictures printed on thick glossy paper and do an admirable job of illustrating the wide range of forms that can be achieved and their spectacular architectural beauty. Recent books by Tony Robbin, Horst Berger, and Kazuo Ishii improve upon these pretty picture efforts by providing insights into the development, design approaches, and detailing technologies of these fascinating structures. But *The Tensioned Fabric Roof* is the first comprehensive publication in this area, and it will be highly useful to students, design professionals, constructors, and potential owner/users.

Craig Huntington was heavily involved in the production of a publication entitled *Tensioned Fabric Structures: A Practical Introduction*, published in 1996 by the American Society of Civil Engineers (ASCE). At the time, I was fortunate enough to be the Chair of the ASCE Special Structures Committee, which produced this document. I got to know Craig quite well and came to appreciate his considerable writing talents, a rarity for both architects and engineers. In the workings of the committee, the paucity of reference materials became very evident. I'm certain Craig's service with this group influenced the direction of *The Tensioned Fabric Roof*.

Craig Huntington covers all the bases about this business, from basic shape descriptions and form-finding, through connection detailing, to fabrication and erection. In each case, he is careful to point out what is different about the design and construction of *these* structures as opposed to conventional "hard" systems. This is of great value to the novice architect or engineer just attempting to learn about tensile membrane design, a subject discussed in only a few schools of architecture or engineering. He draws upon his several years of tutelage in the firm Geiger Berger Associates (renowned in air-supported and tensile membrane design) and upon the resources of his own firm, Huntington Design Associates of Oakland, California. To illustrate principles and explain "how-to" procedures, he frequently utilizes projects from his firm, not out of professional egoism, but because these are designs with which he is logically most familiar. The reader will benefit greatly from this familiarity and from Craig's years of experience with tensioned fabric.

For anyone wishing to get involved with the exciting field of tensile structural art/engineering, this book presents an ideal opportunity for launching such an experience. For the experienced practicing professional, it is an important and valuable addition to his or her library.

R.E. Shaeffer, P.E.
Professor of Architecture
Florida A&M University

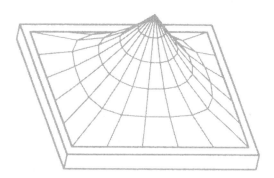

Preface

In 1978, as I neared the completion of my graduate studies in structural engineering at the University of California at Berkeley, I began compiling a stack of magazines that contained articles on what I considered to be the most interesting contemporary structural engineering work: long-span bridges, high-rise buildings, and stadiums. When the stack reached waist height, I began a list of the structural engineering consulting firms that had designed these structures, and with whom I might seek work. The job market for structural engineers was strong at the time, and I received offers from a top firm in Boston and Geiger Berger Associates of New York City, a young partnership that had already completed several innovative stadiums or arenas using what seemed at the time a highly improbable and exotic structural material: tensioned fabric.

Laying out the plusses and minuses of the two offers, the choice appeared easy. Among numerous advantages, I preferred the prospect of living in Boston to New York, and the Boston firm had offered significantly more money. Against all apparent reason, though, I selected Geiger Berger. It was as if I had no real choice in the matter, having already been seduced by the voluptuous forms of tensioned fabric.

My files still contain a copy of the November 10, 1977, *Engineering News Record*, the cover of which showed David Geiger and Horst Berger examining a dramatically curved stretched fabric model of a partially retractable bullfight arena roof. Other roofs may have been larger or more complex than the Geiger Berger roofs completed at that time, but there was a compelling beauty about the fabric roofs, an unsurpassed elegance in their economy of form and material that drew my eye inexorably. I date my love affair with the tensioned fabric roof from that time in 1978. It continues unabated to this day.

My first project at Geiger Berger was the design of the Hajj Terminal at Jeddah International Airport in Saudi Arabia. That roof, approximately 500,000 m² in area, still stands today as the largest tension structure ever built, as well as one of the most innovative. The computer programs used for its shaping and analysis had just been developed, the details of its erection and tensioning were largely untried, and we worked without benefit of relevant building codes, textbooks, or other references. A sizable body of tensioned fabric structure work has followed the completion of Jeddah's Hajj Terminal construction in 1981. The software developed since then is both more accurate and more "friendly," and a vocabulary of connection and tensioning details of proven performance has evolved.

Reference material for prospective designers of tensioned fabric roofs remains extremely limited, however. Until recently, this has been the natural result of technology that was rapidly evolving and known only to a few. With a large body of work now in place, however, the time for more systematic documentation of the state of the art is at hand.

Although a number of books on tensioned fabric structures have come to press over the past several years, most have provided only an overview of their technology and the elements of their design and construction while lavishing most of their words and pictures on case studies of built structures. These books have been welcomed because they provide ample exposure of the many dramatic and eminently photographable buildings that have used fabric in recent years.

The case study books have generally been of limited help, however, to the designers and builders who have sought deeper understanding of how these structures "work," with an eye toward developing their own skills in this most esoteric of building technologies. Accompanying the picture books have been treatises filled with complex differential equations related to the analysis of tensioned structures. These, too, have been of limited utility, as they have addressed only one of the many aspects of a successfully realized tensioned fabric roof. The reader of such a text learns nothing about what shapes may be appropriate for a particular application, how to select an appropriate material, how to make connections between the various elements of the roof, or how the roof may be best fabricated and erected.

I have sought to fill this gap by structuring my own book around the various elements of fabric roof technology and fabric roof performance. Over the course of writing it, I have told friends, only half in jest, that the best prospects for the success of my own volume lay in the fact that those who knew the most about fabric structures were often either too busy or too jealously protective of their knowledge to write about it for the benefit of others. In writing a book that systematically documents the state of the art of tensioned fabric roof technology, it is my hope that the long fabric spans, which must appear to the uninitiated to work by magic, will become accessible to the mainstream of the contemporary building industry, and to all of those who seek to understand new technologies used in the service of utility and beauty.

The enormous varieties of fabric roof forms and design elements do not lend themselves to formulaic approach. They are best illustrated by example, and I have attempted to buttress each theoretical point in the text with the practical evidence from one or more built structures. The structures illustrated and discussed in the following chapters do credit to their creators, who are listed in the Project Credits at the end of the book. They are not necessarily the "best" tensioned fabric structures built; nor, by inference, are those I have left out necessarily less worthy. My selections have been driven by the availability of factual information, drawings, and photographic images, and, above all, by the usefulness of the particular structure in illustrating a specific technical point.

Many of the designs that I have selected as examples are my own, not because I imagine these works superior to those produced by others, but simply because their peculiar nature is well known to me. As Pier Luigi Nervi, the unsurpassed master of reinforced concrete engineering, wrote in *Structures* in 1955, "Structure unveils its nature and the most interesting aspects of its behavior to its creator, designer, and builder" (Nervi 1955).

Some readers will note the use of "he," "him," and "his" throughout the text when the gender of the person referenced is unknown. This deference to traditional language was made in the interest of avoiding either the clumsiness of "his/her" or the confusion resulting from alternating gender reference throughout the text. No slight to women should be inferred.

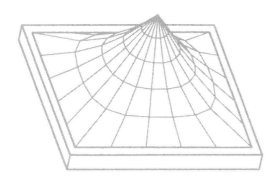

Acknowledgments

This book is dedicated to the memory of my father, Bert G. Huntington, whom I thought of often during the course of my writing. By inviting me to rummage about his own structural engineering office, to watch him work, to play with his drafting tools, and to visit construction sites, he provided the best introduction to the profession that any boy could hope to have. Ten years after his death, I still speak regularly with former clients and employees of his who smile warmly as they recall his generous spirit and engineering wisdom. From him, I came to believe that to design something elegant and useful was the greatest work to which a man could aspire. His spirit animates my own desire to create beautiful tensioned fabric structures and to share the secrets of this medium with others.

I signed a contract to write a book on tensioned fabric structures more than ten years ago. Since then, each of the countless hours that I spent at my home office reviewing references or editing drafts was an hour that I was unable to spend with my wife Robin Huntington or my daughter Samantha Huntington. They indulged my absence from the more domestic regions of the house with extraordinary generosity. In compensation, I can offer them only heartfelt thanks for allowing me the opportunity to complete a project that I found immensely satisfying and that I hope makes a small contribution to the quality of our built environment.

As my father believed that I could be a successful structural designer long before there was evidence to support such an opinion, so Signe Hammer believed that I could be a successful writer long before I believed so myself. Twenty years after completing her nonfiction writing class at New York City College, I look back on her critiques of my writing and her relentless verbal sparring with me, on topics ranging from abortion rights to classic architecture, as formative experiences in developing my own skill with language.

Every published work requires the positive input of many people, and this volume is no exception. I am indebted to Joy Chau, the editor who first encouraged me to write a book on fabric structures for ASCE Press. Her successor, Bernadette Capelle, along with Book Production Manager Suzanne Coladonato, guided the book ably to its completion. Nearly all of the CAD drawings were completed under the able mouse guidance of Nancy White. Photographs were generously supplied from many sources, including Pat Nowicki at Birdair, Marcel Dery at Chemfab, Professor Jörg Schlaich of the University of Stuttgart, Don Douglas of Sullivan & Brampton, Enrique Limosner of San Francisco City College, Charles Duvall, and Ron Shaeffer of Florida Agricultural and Mechanical University. Professor Shaeffer also provided innumerable valuable suggestions on the manuscript in his role as technical reviewer for ASCE Press.

The portfolio of color photography in this book would not be present without generous financial support from the fabric structures industry. The contributions of Birdair, Sullivan & Brampton, and Eide Industries have made this a better and more enjoyable book.

Comprehensive knowledge of tensioned fabric structures requires competency in a broad range of fields, including materials, architecture, structural analysis and design, fire safety, acoustics, and mechanical and lighting design. I am not expert in all of these fields and have therefore relied on the information and feedback provided by various specialists. Valuable input regarding materials was provided by John Effenberger and Marcel Dery of Chemical Fabrics Corporation. Mr. Dery also reviewed a draft of Chapter 4. Steve Goodson, a colleague from my years at Geiger Berger, reviewed and helped me to solidify Chapter 5. Alexander Itsekson, formerly of Huntington Design Associates and now owner of his own firm, Engineous Structures, provided many useful comments on an early draft of the book.

Over the past 20 years, I have had the opportunity to work with and learn from some of the world's most experienced, knowledgeable, and innovative tensioned fabric structure designers. These men include Horst Berger, David Geiger, Jim Ford, Steve Goodson, and Chris Anastos. Without their generous nurturance, I would have no knowledge worth sharing about this topic.

The Tensioned Fabric Roof

At the Hajj Terminal at Jeddah International Airport in Saudi Arabia, 210 matching modules of sunbaked white fabric cover up to 35,000 people at a time under what may be the world's largest roof (shown on the next page). In Riyadh, 300 km to the west, another creation of white fiberglass fabric shields the grandstands from the hot sun around the 900 m perimeter of the grand new sports stadium (shown in Chapter 3). In less dramatic form, the same material is used in dozens of shopping malls across the United States to provide light-filled and festive atrium spaces.

What is a tensioned fabric structure? The term describes what a layman might call a "tent," a word those in the industry shun out of distaste for its connotations of flammability, impermanence, and dusty nomadic peoples. More to the point, tensioned fabric structures are covers or enclosures in which fabric is preshaped and pretensioned to provide a shape that is stable under environmental loads.

Such a dry definition, however, fails to capture the enduring nature of their attraction: tensioned fabric structures are shamelessly and voluptuously curved shapes in a world of right angles. They are high-tech forms that boldly display the machinery of their construction in an architectural environment preoccupied with the artful use of fascia and finishes to hide all traces of a building's working.

A Layman's Fascination and an Architect's Perplexity

The appeal of tensioned fabric structures to the man on the street is immediate and visceral. He is excited by their dramatic and light-filled forms in a way that he often cannot explain. The architectural

The Hajj Terminal at Jeddah International Airport in Saudi Arabia is the world's largest tensioned fabric roof.

Source: Birdair, Inc.; used with permission.

press, whose job it is to provide such explanation, seems unable or unwilling to do so, as if the words they use to describe the visual vocabulary of contemporary architecture are unable to describe the more exotic forms of fabric. Enchanted by the engineering methodology by which they are made to work, however, editors find an uneasy place for fabric structures in the technology pages at the backs of their magazines.

It is the purpose of this book to bridge the gap between the layman's fascination with the dramatic forms and expressive details of tensioned fabric roofs and the leading-edge technology by which they are designed and constructed, and then to provide some guidance to the designer or builder who would bring one to reality.

Tensioned fabric structures have a kinship with the best shell structures of concrete or stone in the way that their forms are derived largely from considerations of structure and reflect, in highly expressive form, the flow of forces within the structure. Unlike more conventional building forms, their structural design is integral to their

[LEFT] Jörg Schlaich's Stuttgart Garden Fair roof achieves a structurally expressive form with a glass fiber-reinforced concrete shell of only 10-mm thickness.

Source: Jörge Schlaich; used with permission.

architectural expression, and the usually sharp demarcation between the roles of the structural engineer and the architect becomes blurred. The effect of this on the process of design and construction is described in detail in Chapter 9.

For reasons both practical and cultural, the vast majority of contemporary construction is rectilinear, its forms described by straight lines and right angles. Odd angles and curving forms (generally circular) are used sparingly and, usually, for dramatic flourish in contemporary architecture. Their sometimes powerful effect is seen in the East Building of the National Gallery of Art in Washington, D.C. Designed by the architect I.M. Pei, the building's sculptural form features stark exterior walls that come to a knife-edge prow at one end. So startling is the effect of this acute angle that the white stone fascia at the corner was stained black from the handprints of visitors soon after it was opened to the public.

In tensioned fabric roofs, however, nearly all forms are curved forms and few angles are right angles. Architects only infrequently find a role for them in the restrained vocabulary of contemporary design. Typically, they are relegated to a minor dramatic role as skylights, or used in amphitheaters, sports arenas, or other recreational facilities where their splashy shapes can be more readily accepted.

The fabric roof's visual misfit with the mainstream of contemporary architecture has been partly responsible for its neglect by the architectural press. There are other reasons, though, rooted in the organic, structurally derived nature of their form and detail, that play an equally important role. The vocabulary of architectural forms has always been filled with references to religion, the natural world, geometry, and architecture's own history. The modern movement in architecture dispensed with much of this, but the careful massing of

[ABOVE] Clarity of form and economy in material use are hallmarks of the concrete shells designed by Pier Luigi Nervi, consultant on St. Mary's Cathedral in San Francisco.

Aside from their use of fabric, traditional circus tents have little in common with contemporary tensioned fabric roofs.

Source: Canobbio S.p.A.; used with permission.

forms and the visual patterning of building facades were filled with direct references to the work being created contemporaneously by artists and sculptors. The history of architecture (and the history of architectural writing) follows in lockstep with the history of culture.

Much of the connection between built form and culture is lost with tensioned fabric structures, concrete shell roofs, and other structurally derived forms, except, perhaps, for the manner in which they and the culture celebrate technology itself. Their shapes reflect no religious motif or abstract geometry as much as the path that their loads travel in reaching the ground, just as the form of a tree embodies the distribution of the forces of wind and self-weight from its peak down to the ground. The design of the connections between the members that make up a fabric roof recalls nothing beyond the visually expressive transfer of tension or compression forces between structural elements.

The tensioned fabric structure, then, is stripped of much of the cultural underpinning that provides the basis of architectural writ-

ing. Lacking reference to time and place, this architecture of ancient nomadic peoples has found no home in the chronicles of contemporary architecture.

If tensioned fabric roofs lack reference to either historic architecture or to other contemporary art forms, however, they are replete with reference to their own past as tents. Military tents, circus tents, and nomadic tents all comprise the context in which we view today's computer-generated shapes of gleaming synthetic fabric. The references do not warm the hearts of industry marketing executives, who find that comparison to structures that were generally cheap, temporary, unreliable, dirty, and flammable do not enhance their ability to sell a contemporary product that is often none of these things.

Where the dictates of high culture are lost, however, the laws of physics governing structural behavior and the limitations of available technology remain as guides to form and detail. There are voluminous writings about tensioned fabric structures devoted exclusively to their technology: the behavior and chemistry of their

[ABOVE] The stone ribs and ceiling of a Gothic cathedral such as St. Etienne predict the shapes of steel pipe arches and fabric membrane that would follow seven centuries later.

Source: Enrique Limosner; used with permission.

[ABOVE LEFT] The fabric roof of the Walden Galleria in Buffalo, New York, brings an abundance of light to the traditional vault.

Source: Birdair, Inc.; used with permission.

[RIGHT] Charles Duvall's canopy provides a dramatic entry to Carleton Centre in Ottawa, Ontario.

Source: Charles Duvall; used with permission.

[ABOVE] By placing diagonal bracing on the outside of the building, Skidmore Owings & Merrill structural engineer Fazlur Khan made the structure of Chicago's John Hancock Building an integral part of its architectural expression.

Source: Enrique Limosner; used with permission.

materials, the mathematical description of their shape, and determination of their internal forces. In writing this book, I seek to bridge the gap between these technological treatises and conventional architectural writing. This will not be done through the linking of built form to culture, as these connections are not profound in tensioned fabric structures, and the occasional analogies made between their shapes and forms in nature seem rather beside the point. The excitement of fabric roofs lies in the bold expression of emergent technologies and the satisfaction of the dictates of physics to cover space in an efficient and clearly expressed manner. It is this quality that makes the layman's eyes widen when he sees a long, free span of fabric, and it is this quality that I try to bring some understanding to.

Varying Priorities in the Design and Appreciation of Tensioned Fabric Structures

In the layman's view, the work of architects is associated with the creation of beauty and the work of engineers with the satisfaction of practical needs. Clearly, these generalizations miss the mark. The architect, as a design generalist, must address the needs of a building's occupants for shelter and safety, at the same time that he seeks to provide visual delight. Likewise, the great tradition of beauty established by engineers such as John Roebling, Gustave Eiffel, Robert Maillart, Pier Luigi Nervi, and Fazlur Khan demonstrates clearly how successful the best structural engineers have been in achieving a character of beauty that is noteworthy for simplicity, clarity, and economy of form.

[ABOVE] The rigging that supports El Grande Bigo is perfectly at home on the working waterfront of Genoa, Italy.

Source: Canobbio S.p.A.; used with permission.

It is not so much that architects are creators of beauty and engineers providers of utility, then, as that the differing training and outlook of architects and engineers are emblematic of differing priorities in design. Specific reference to tensioned fabric structures helps to illuminate the differences.

Curvature is a requirement for the stability of a fabric structure form. This characteristic gives them inherent drama in a world dominated by the straight lines and rectilinear forms of modern construction. To the imaginative eye of an architect, their shapes may evoke images of eggshells, seashells, breasts, sails, mountains, bubbles, or historic architectural forms. The architect Anthony Belluschi, who has occasionally used fabric in his designs, offered a parallel between the vaulted ceiling of a Gothic cathedral and the crossed arch form of a fabric-covered shopping mall atrium roof.

[ABOVE] The surprising strength and stiffness of thin-shell construction finds a natural parallel in the common eggshell.

Source: Ronald Schaeffer; used with permission.

An architect might use supporting cables and masts, whose straight lines provide a counterpoint to the curves of the fabric, to recall industrial rigging or derricks. Similarly, a long pointed mast that extends high above the fabric it supports, its tip punched aggressively to the sky, may evoke the image of the pole of a battle flag.

When not finding such historical or symbolic parallels, architects often appreciate fabric roofs for their purely sculptural appeal. Through shape, color, or the light-transmitting quality of their surfaces, fabric roofs bring drama to every building in which they are applied, and provide an architectural focal point when used as a small part of a larger architectural statement. The structural aspect of tensioned fabric is abandoned entirely in favor of sculpture when the material (often an unpatterned, stretchy knit) is used for dramatic flourish in the interior of a building.

The sophisticated Electronic Arts awning provides partial shading of direct sunlight to the inside of the window wall [ABOVE RIGHT] and partial protection from the weather to the outside of it [ABOVE LEFT].

A frequent architectural counterpoint to the drama of tensioned fabric roofs is their playfulness. They move sympathetically with the wind and, seen from different viewpoints, their shapes are constantly changing. Some designs are actively coquettish, with fabric cutouts or lacy cabling that tease the viewer with simultaneous images of light and shadow, inside and outside, enclosure and exposure.

An architect also may take advantage of the lightness and translucence of fabric to soften the transition from outside to inside at the entrance to a building. While fabric awnings are a staple of traditional Main Street storefronts, their design becomes high architecture in buildings such as Electronic Arts in Redwood City, California.

To an aesthetically sensitive structural engineer, though, the best tensioned fabric structures might be designed and appreciated with the same sensibility as a bridge by Maillart or a concrete shell conceived by Pier Luigi Nervi. In this "structural aesthetic," designs are admired not so much for symbolic imagery or sculptural qualities as for the raw beauty associated with economy of material use and clear expression of load-carrying mechanisms and the interplay of tensile, compressive, and bending forces. David Billington, our most prolific chronicler of beautiful bridges and building structures, went so far as to coin the term "structural art" for such work (Billington 1983).

Clearly, the broad generalizations given here do not define the aesthetic proclivities of all architects and structural engineers. Some architects have an appreciation of structural behavior and seek to express it in their designs. By the same token, many structural engineers appreciate the broader aesthetic goals of the architect and work in concert with him to achieve them.

[LEFT] The cat's-eye cutouts of the fabric for my design of the Weber Point Events Center in Stockton, California, give those beneath the canopy a glimpse of the sky above while allowing bright beams of direct sunlight to break the soft shadows on the grass below.

Source: Don Douglas; used with permission.

[ABOVE] White fabric sails create a playful focal point in the Cincinnati Bell Telephone interior.

Source: Charles Duvall; used with permission.

In most construction mediums, aesthetic choices fall almost entirely to the architect. In tensioned fabric structures, however, form and structure are the same, and only a moderate degree of shape manipulation is possible while maintaining structural stability and load-carrying ability. Likewise, structural connections and details are typically left exposed. In tensioned fabric structures, therefore, the visual impact of the structural engineer's work is enormous, and some degree of structural expression becomes inevitable. In fabric structures, collaboration between architectural and structural designers is of paramount importance, and successful designs represent an appropriate marriage of the aesthetic and practical goals of both disciplines.

2 Elements of Form and Design

No structural type—whether it uses steel, wood, concrete masonry, concrete, or some more exotic material—conveys such an image of freedom of form as the tensioned fabric structure. This image is at once both fabric's best selling point and the bane of fabric structure engineers, as closer examination of fabric structure design reveals that the dramatic and voluptuous forms of fabric roofs respond to strict rules of form and shaping. Also, the properties of structural fabrics limit the range of shapes available in fabric, in some ways more so than those using more conventional materials (Huntington 1984).

One thing that fabric roofs can never be, unfortunately, is the thing that they are so frequently called: tension structures. Compression members are structural elements whose strength (and resultant efficiency) is limited by their tendency to buckle under heavy load, particularly when they are slender in cross section. Tension members, on the other hand, can be stressed all the way to the yielding point of the material before failing, no matter their slenderness. Noting the inherent efficiency of tension members, some have suggested that "tension structures"—composed entirely of tension members—represent the ideal of structural efficiency, and that fabric (having high tensile strength and negligible compression strength) is the ideal construction material.

In practice, however, tension structures are no more common (and no more attainable) than perpetual motion machines. Intuition alone indicates that the downward pressure exerted by the weight of a structure (as well as the additional downward pressure of live loads acting upon it) must be resisted by the upward resisting pressure of one or more compression members. Close examination of any presumed tension structure reveals the presence of compression members. In a suspension or cable-stayed bridge, for example, the tensile force in the cables that support the bridge deck is reacted by the compressive force in the towers. Fabric tension structures utilize cables that are in tension throughout their length and membranes that are in tension throughout their surface, but these tensile forces are inevitably reacted by masts, arches, or similar compression members. Air-supported roofs appear to offer an exception, until we recognize the pressurized air itself as a compression member (Huntington 1989).

The structural efficiency and elegance of fabric roofs do not lie in their presumed tension structure behavior but in the fact that their membranes are "form resistant": relying on their curved shapes to resist load and carry it to supporting compression elements in pure tension, without the less efficient bending resistance relied upon by beams, slabs, and other more conventional structural members. This chapter explores the fundamentals of tensioned fabric membrane behavior and how designers exploit them to create efficient and beautiful structural forms.

Principles of Design

By understanding a few simple principles of fabric roof behavior, engineers, architects, and others can begin to understand some of the constraints on their design and conceive ways of achieving key design goals with structures that are stable and economical.

It is easy to comprehend the way that a fabric roof carries load by first considering it as a grid of individual fabric threads. A single thread, primarily because of its extreme slenderness, has considerable strength when pulled from the ends (placed in tension), but negligible strength when either pushed on the ends (compressed) or loaded transversely at some point along its length (bent) (see [a] in the drawing on the next page). Conventional structural materials such as steel, reinforced concrete, and wood, by contrast, are typically configured with cross sections that are stout enough that a single linear element (analogous to a thread) has significant resistance to tension, compression, and bending loads (see [b] in the drawing on the next page).

Extrapolating these ideas into a woven fabric material, we see that a flat sheet of fabric is also strong when pulled along opposing edges (placed in tension) but is weak when either compressed or bent. Taking advantage of the tension strength of fabric while circumventing its weakness in both compression and bending is the primary motive underlying the principles of shaping tensioned fabric structures.

Sometimes, designers add steel or other "conventional" elements to provide compression and bending resistance. They also

[a]

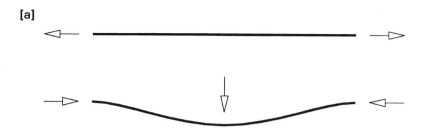

[b]

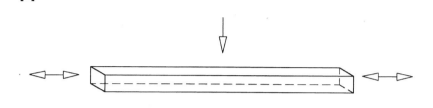

[c]

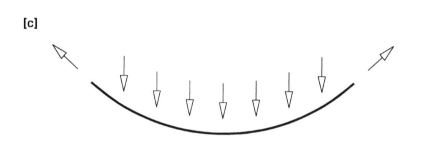

(a) A thread of fabric is strong and stable under tension (top) but must change its geometry to resist either compressive or out-of-plane forces (bottom).

(b) Steel, reinforced concrete, timber, and other "conventional" materials are typically stout enough to resist tension, compression, and out-of-plane forces.

(c) Curvature provides resistance to out-of-plane forces.

provide such resistance through the introduction of curvature and pretensioning into the fabric itself. In the same manner that curvature in the suspension cables that support a bridge deck provides resistance to the vertical loads acting perpendicular to the bridge deck, the curvature in fabric provides resistance to the wind or live loads that act out of the plane of the fabric (see [c] in the drawing above) (Huntington 1995).

Because the weight of a bridge deck is usually in excess of any upward suction due to wind, a bridge suspension cable needs to resist only downward loads. A fabric roof, however, has a large exposed "sail" area and minimal weight; therefore, it must generally resist wind loads that both push inward and cause suction outward on the fabric, in addition to dead and live loads that act vertically downward. The fabric roofs that curve outward at all points in dome-like form are termed "synclastic." Suction forces acting on them are resisted by an increase in the tension in the fabric, while downward loads are resisted by internal air pressure in the building enclosure (see the pair of drawings at the top of the next page). These "air-supported" roofs are used both for high pitched "tennis bubbles" and the more elegant forms of low-profile stadium roofs.

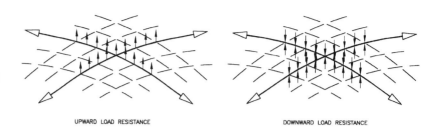

UPWARD LOAD RESISTANCE   DOWNWARD LOAD RESISTANCE

[RIGHT] On a synclastic surface, upward loads are resisted by a stress increase about both axes of the fabric, while downward loads are neutralized by internal air pressure.

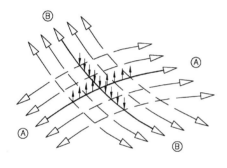

[ABOVE] On an anticlastic surface, fibers with convex curvature, A, increase their tension to resist upward loads, while those with concave curvature, B, increase their tension to resist downward loads.

More commonly, though, fabric roofs are made stable by providing curvatures both inward and outward at all points of the fabric surface. In practice, this is achieved by curving the fabric fibers about one axis of the fabric in a convex manner and those about the other axis in a concave manner. Fabric or other surfaces having such opposing curvature are said to have "anticlastic" shapes (see the single drawing below left) (Huntington 1995). As the predominant shapes in fabric tension structure design, anticlastic forms are the focus of this book in general, and of the discussion that follows.

Generating Anticlastic Shapes

Lacking pressurized air support, no fabric roof of significant size can safely resist varying loads without anticlastic curvature. However, the designer who is able to visualize and manipulate anticlastic forms gains access to the range of wildly dramatic to subtly elegant forms possible in tensioned fabric roofs. This range of forms is achieved by manipulating the varying elements that support or restrain the fabric, elements that include cables, arches, masts, trusses, and ring beams. The roof forms that result can generally be classified either as saddles or cones, and these two types can be manipulated and combined to create the entire range of fabric roof shapes.

In the saddle, anticlastic shapes are generated by point, linear, or curving perimeter supporting elements that utilize elevation variations to generate the doubly curved surface (see the drawing at the top of the next page) (Huntington 1995). Combinations of saddle shapes are created by the use of arches or similar curving elements to subdivide a given perimeter shape (see the drawing on the right side of the facing page) (Huntington 1995).

In a saddle, fabric curvature is created by the use of perimeter supports (point, linear, or curving) that do not all lie within the same plane. In a cone, on the other hand, the perimeter supports may all lie within one plane, but curvature is induced by connecting the fabric to a point inside the perimeter that is out of the plane of the perimeter supports. Cones or inverted cone shapes result (see [a] and [b] on page 16) (Huntington 1995). The configuration of the

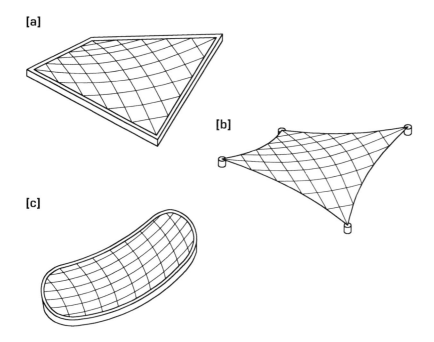

[LEFT] Three basic saddle shapes: (a) linear boundary, (b) catenary boundary, and (c) curving boundary.

perimeter supports may be varied and additional point supports may be added in order to vary the shape of conical roofs, or cones may be used in combination, in the same manner as saddles (see [c] and [d] on page 16).

Fabric roofs may more accurately be described as "tensioned fabric structures" than the more common "fabric tension structures." The compression members most commonly used to support tensioned fabric are masts, the linear compression elements used in cone structures, and arches, the curving compression elements that shape saddle roofs. Masted structures tend toward forms that are pointed, outreaching, and dramatic, while arched structures tend toward forms that are rounded, sheltering, and subtle.

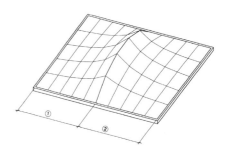

[ABOVE] A simple arch-supported shape is a combination of two saddles, each having three linear edges and a single curving edge (the arch).

The Art of Design

In tensioned fabric roofs, the art of creating form lies in the skillful manipulation of support conditions to create shapes that are aesthetically and functionally satisfactory, economical, adequate in load-carrying ability, and reliable. There are a handful of pitfalls that commonly make a design concept difficult or impossible to realize. These are described below.

Inadequate Fabric Curvature

The additional stress that occurs in fabric under load is inversely proportional to the curvature in the deflected fabric shape (see the drawing on page 17). Hence flat shapes with large curvature radii in the fabric are inherently less resistant to imposed loads. The large deflections that fabric structures (especially flat ones) experience under load tend to increase curvatures where required, in a manner

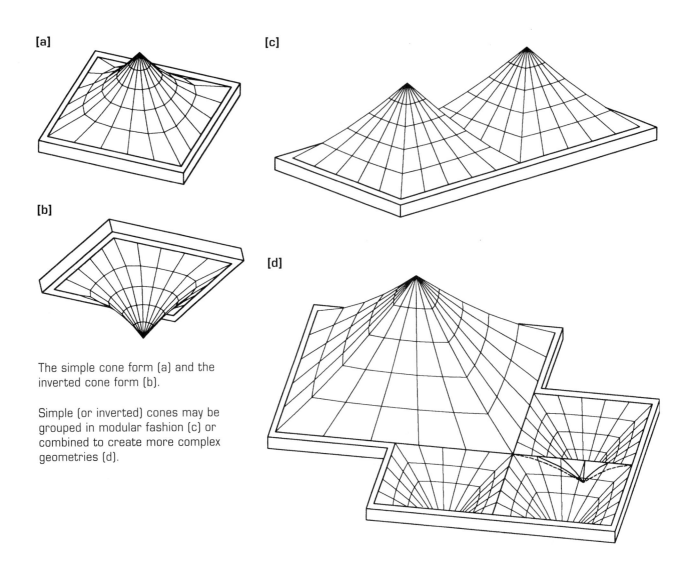

The simple cone form (a) and the inverted cone form (b).

Simple (or inverted) cones may be grouped in modular fashion (c) or combined to create more complex geometries (d).

that improves their ability to resist a given load. This inanimate "intelligence" is seen most clearly in the manner that a completely flat area of fabric sags until out-of-plane loads are equilibrated. This is an inefficient load-bearing mechanism, however, and fabric shapes with little or no curvature under prestress are generally most practical on small canopies with the fabric span limited to 10 m or less. On larger spans, designers usually seek span-to-sag ratios of less than 15, with lower ratios or cable reinforcement required as spans increase further.

Inadequate Cable Curvature

Like fabric, cables are sensitive to changes in curvature. It is therefore impractical to design a fabric roof where the "catenary" cables along the free edge of a fabric roof pass straight from support to support without reasonable curvature. The curvature in ridge, valley, and other cables that lie in the interior of the fabric roof varies with the curvature in the fabric itself and must be considered in conjunction with fabric curvature. The span-to-sag ratio of cables is usually

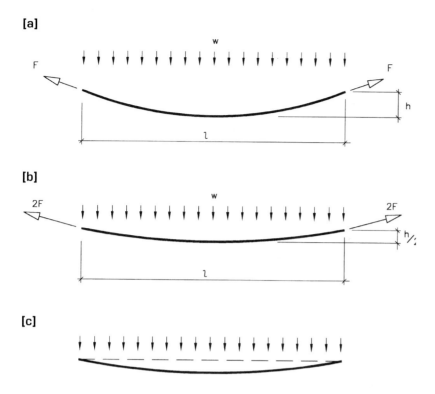

The tension in a cable or strip of fabric is a function of sag length and load.

(a) For typical shallow sags, the tension in a cable or a strip of fabric may be approximated by the formula $F = wl^2/8h$.

(b) Halving the sag h doubles the tension in the element.

(c) Fabric or cables that are constructed without initial curvature are able to carry load only by elongating until they acquire significant sag. Such elements may be subject to excessive flutter under varying load.

limited to 15 on all but the smallest spans, with considerable economy associated with lower ratios.

Inadequate Mast Termination

In the interest of achieving the simplest and most direct fabric roof form, it is inviting to try to terminate the fabric in a point at the top of a supporting mast. However, unless a number of radiating cables are added to the structure to help distribute load into the mast top, the fabric must be terminated in a ring that is large enough to transfer the load from the entire fabric roof into the top of the mast without overstressing the fabric (Huntington 1992). Contemporary computerized analysis techniques provide a ready means of analyzing fabric stresses and determining ring geometry, but simple hand analysis techniques aid in understanding the design parameters (see the drawing on page 18).

Assuming, for simplicity, that the fabric is weightless and that no load is transferred from the fabric to radial cables underneath it, the satisfaction of statics requires that the vertical component of force in the pretensioned fabric be the same at both the base of the fabric and its peak. For this to be so, and for the radial prestress in the fabric to be constant from bottom to top, the circumference at the peak must be apportioned to the circumference at the base in accordance with the ratio of the sine of the angle of the fabric to the horizontal at the peak and at the base.

If radial fabric prestress is force per unit width, f_r, and if the tent is divided into pie-shaped strips of width W_t at the top and W_b at the

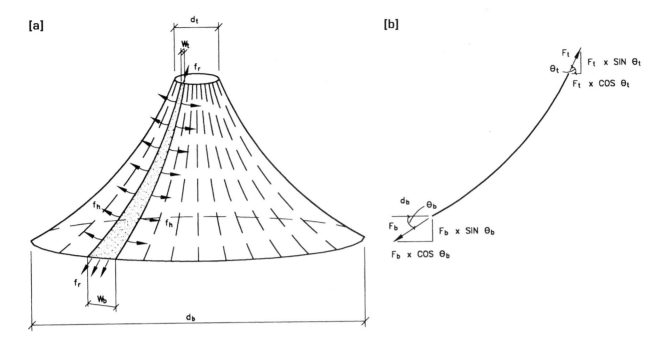

[ABOVE] (a) Forces acting on the pre-stressed strip. (b) Force components at top and bottom of the strip.

bottom, the total forces on each strip, at top and bottom, respectively, are:

$$F_t = f_r \times W_t \text{ and } F_b = f_r \times W_b$$

For the strip to be in equilibrium with uniform prestress throughout its surface, the vertical forces at top and bottom must equate. Therefore,

$$F_t \times \sin \theta_t = F_b \times \sin \theta_b$$

Considering the entire membrane surface,

$$d_t \times \sin \theta_t = d_b \times \sin \theta_b$$

and the required diameter of the top tension ring is determined by the formula

$$d_t = d_b(\sin \theta_b / \sin \theta_t)$$

Where the designer requires a smaller top ring (reduced d_t), this may be accomplished either by allowing higher radial prestress, f_r, at the top than at the bottom, or by adding radial cables whose tension force supplements f_r at the top.

The horizontal load applied to the bottom compression ring may be much larger than that applied at the top tension ring,

$$W_b \times \cos \theta_b > W_t \times \cos \theta_t$$

and equilibrium is established by the inward component of force from the circumferential or "hoop" stress in the fabric, f_h.

As an example, if fabric slope is 10 degrees at the base and 80 degrees at the peak, the circumference of the tension ring at the peak must be 18 percent of that at the base. In reality, some of the fabric stress may transfer into radial cables, the fabric may be reinforced in the region of the peak, or the designer may allow greater prestress at the peak than at the base, all of which will permit a smaller tension ring.

Excessive Aspect Ratio

Designing a roof with a large height difference between its peak and the perimeter supports (resulting in a high "aspect ratio" between the height of the roof and its plan dimensions) tends to generate plenty of fabric curvature, but it also increases the exposure of the roof to lateral wind loads, increases the area of fabric that must be used in order to cover a given plan area, increases the volume of air that must be temperature controlled, and results in excessively long compression members. An architect's desire for a high-profile shape with dramatic and voluptuous curves must be balanced against these considerations.

Long Compression Members

Long masts, arches, or other compression members are prone to buckling and require large and heavy cross sections in order to achieve adequate capacity. Excessive compression member length is generally related to excessive aspect ratio (described above).

Reversals in Curvature

With certain supporting structure geometries, inadequate fabric curvature may occur when membranes with generally appropriate curvature radii are forced by supporting structure geometry to pass through reversals of curvature. The radius of curvature reaches infinity at these inflections, and the resulting flat area may be overstressed or may deflect excessively under load.

Unstable Supporting Structure

The most expensive and durable fabrics currently available use a fiberglass substrate that has high tensile strength, but is thin, damageable in handling, and limited in tear resistance. Small tears caused by overstress or handling damage have propagated into lengthy rips in the fabric of a number of roofs. Until fabrics with near perfect reliability are developed, structures with moderate to long spans and supporting members of considerable weight must not be reliant on the fabric for stabilization of the masts, arches, or other rigid members. Rigid supporting member base connections can provide stability, though typically at significant expense. In general, therefore, good designs make use of either cable nets in the plane of the fabric (seen most often in European practice), guy cables connecting the top of the mast or arch down to grade, or crossed arch or A-frame designs in which the supporting elements are self-bracing.

The above list is both general and incomplete. Through awareness of these potential design pitfalls, however, in combination with careful observation of successful built structures and consultation with an experienced fabric roof engineer, the architect or other designer can begin to conceive fabric roof forms that achieve aesthetic and functional excellence with forms that are safe, reliable, and economical.

3
Tensioned Fabric Structural Systems

Chapter 2 outlines the basic principles of tensioned fabric roof behavior and describes the basic building blocks—cone and saddle forms—that are used to create fabric roofs. In practice, designers often use repeated modules of a particular form or combine shapes of varying geometry, even wedding cone and saddle shapes in the same structure. The parameters of fabric roof design are broad and the freedom to create exciting and evocative shapes is great, so long as the designer avoids pitfalls such as those described at the end of the previous chapter.

Variability of form is a hallmark of tensioned fabric construction, and roofs can be adapted to fit a wide range of building footprints. Because of the curvature requirements of the membrane, however, tensioned fabric structures typically have fairly tall profiles in elevation and cannot easily be adapted to the flat roof profiles characteristic of conventional construction. Flat fabric profiles are only possible on small structures or those with a tightly moduled supporting structure (such as a space frame or cable net) that supports the fabric at close intervals. In the latter type of application, the fabric is more a cladding material than a true structural element.

An attractive feature of tensioned fabric structures is their enormous range of spanning capability. Membranes have been used in countless applications as an alternative to translucent glazing over spans of 3 to 20 m or more. Fabric has been applied just as effectively in stadiums and other assembly structures with spans of more than 200 m. In these applications, the fabric is typically restrained or supported by steel cabling in conjunction with air support or rigid steel elements, so that the unsupported span of the fabric itself is seldom greater than 30 m. In these roofs, the fabric provides a significant portion of the strength and stiffness and is integral to its global behav-

[RIGHT] The articulation of cable stays and mast-to-cable connections makes the Folkestone Chunnel Terminal in London an elegant application of the king post system.

Source: QA Photos Ltd.; used with permission.

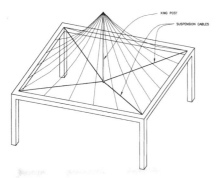

[ABOVE] King post designs eliminate the conventional cone's need for internal column support.

ior. Good design of these structures explicitly considers the properties of the fabric, and these roofs are therefore appropriately considered true tensioned fabric structures rather than fabric-clad roofs.

Both air-supported and cable dome roofs can and have been sheathed in materials other than fabric, including steel, plexiglass, and prestressed concrete. In general, though, the adaptability of fabrics to the deformations in flexible cables and the ability of fabrics to wrap over cables without direct attachment have contributed to their unsurpassed economy in cable-supported or cable-restrained structures.

The generation of fabric roof forms primarily involves manipulating the geometry of the masts, arches, ring beams, cables, and other boundary elements with which the fabric must interface. Commonly used structural systems are described in the sections that follow.

The categories given are arbitrary to a degree, and large overlap occurs between certain systems. Suspended roofs and cantilevers might be seen as permutations of the cone, for example, while cable domes are a sophisticated three-dimensional variation of the arch. Air-supported roofs and air-inflated lenses are close cousins. In the latter, pressurized air is held between two skins of fabric, while in the former the pressure vessel is expanded to encompass the entire building.

Cones

Mast-supported cones are the most basic of tensioned fabric roof forms and also provide the greatest possible range of shapes including single peaks, a near infinite range of multiple peak configurations, inverted cones, and those with fixed or catenary edges. Their popular-

ity stems from a variety of advantageous features. Not least in importance is the familiarity of their form. The traditional peaked shape of the circus tent has been the visual precedent for countless party rental tents, amphitheater covers, and other entertainment applications.

Cones have their drawbacks, as well, beginning with the fact that, in their classic mast-supported configuration, posts must be founded in the interior of the floor space, disrupting possible free spans. King post designs, with the base of the mast terminated above the floor and supported by cables rising to the perimeter, eliminate this potential disadvantage (see the photographs on the previous page). Other more complicated configurations, including the suspended designs discussed later in this chapter, use alternative means of retaining peaked cone configurations without interior support.

Cones are also prone toward high fabric utilization relative to their plan area, particularly in single-peaked configurations, due to steep slopes near the peak. They may also have higher energy use related to the large volume of conditioned area associated with high peaks.

While cluttered with the extravagant ornamentation of the gaming environment, Harrah's Carnaval Court illustrates the basic conical roof form.

Source: Don Douglas; used with permission.

The twin-peaked form of the amphitheater constructed for National Semiconductor is integrated into a park-like corporate setting.

Source: Bernard André; used with permission.

A few examples exhibit the wide range of cone applications. In their most basic forms, they have a symmetrical plan shape with a single center peak supported by a vertical mast. My firm engineered such a structure at the Harrah's Casino in Las Vegas, Nevada (see photograph on previous page). The structure was conceived by the architect to have an overall shape and striped color bands that evoke the image of an old-fashioned carousel. While it has been cluttered by signs, banners, lights, sculptures, and the other accoutrements of the gaming environment, the basic form is simple: a 12-sided polygon, 13 m on a side, with catenary edge cables and a perimeter pipe compression ring. The fabric rises upward to terminate at an inside cable hoop 9 m in diameter that is supported from the center mast by 12 suspension cables. The center mast is not an obstruction here but an advantage: it is encircled by a large, sculptural bench that is the focal point of an open courtyard between several casino structures.

The structure has one deviation from pure symmetry. Primary pedestrian access to the area under the canopy is from two opposite sides. A column was eliminated on each side of the structure in order to provide a broader entry. The inward and upward pull of the catenaries coming to the two unsupported corners puts a heavy bending and torsional demand on these two "kinked beams." Using the excellent buckling and torsional resistance of steel pipe on these mem-

bers, however, we were able to make the structurally clumsy configuration work without undue increases in member size.

Most applications do not adapt so readily to fully symmetrical single-masted structures. In amphitheaters, for example, the presence of the action on stage (as well as the most desirable seats) near the centerline of the structure prevents placement of a mast there. Many or most such applications therefore typically utilize a mast at each side of the stage house. This configuration was employed in a structure conceived by architect Bellagio Associates and engineered by my firm for an open-air theater built for National Semiconductor Corporation in Sunnyvale, California (see photograph on the previous page). Like Harrah's, this structure employs catenary cables at the perimeter fabric termination. There is a critical difference in the design of the perimeters of the two structures, however. At Harrah's, the inward pull of the catenaries is resisted by the 12-sided steel pipe compression ring. At National Semiconductor, however, two steel tie-back cables extend outward from the top of each of eight perimeter posts down to grade. The tension in the radial and catenary cables of the roof is in essence carried straight to grade in the tension of the tie-backs rather than being resisted in the compression of the ring. While the ground-level obstruction of tie-backs was unacceptable at Harrah's, the tie-back cabling facilitates a lighter and more structurally elegant solution at National Semiconductor.

The National Semiconductor amphitheater form provides an appropriate and economical solution in an application where the membrane covers both stage and audience. The Weber Point Events Center (shown above) in Stockton, California, demonstrates a different approach. Here, the roof canopy covers only the stage and a small portion of a large grass field that may be filled with spectators

The Weber Point Events Center opens broadly in the front to invite the view of spectators seated on the grass beyond.

Source: Don Douglas; used with permission.

THE TENSIONED FABRIC ROOF

The entrance canopy for National Semiconductor was designed with simple but strictly followed geometries.

Source: Don Douglas; used with permission.

for major events. Supporting the front edge of the membrane on perimeter posts would have obstructed much of the audience's view toward the stage, a problem that we solved by using two pairs of interior masts and sloping the front ones steeply forward. The front of the canopy is thereby held in position by masts founded far back along the sides of the stage.

In selecting a roof form such as that at Weber Point, with masts sloping at varying angles and an overall form that is more abstract than geometric, the designer should beware the added complexity in analysis, detailing, and fabrication that may result. There may be a visual cacophony to structures with multiple sloping masts, as well. While vertical masts are parallel to one another from any view, the angle of sloped masts both to each other and to the horizon varies from every viewpoint, and a design that appears well ordered in a front or side elevation may appear chaotic from oblique angles.

Our office engineered yet another type of cone in a small entrance canopy built for National Semiconductor (shown above). Here, the relatively free form of the amphitheater is replaced by a more rigid geometry of straight lines and circular curves. While strict geometries can lead to structural (and cost) inefficiencies in large roofs, they are of little consequence in structures the size of the National Semiconductor entrance canopy (9.7 m overall length) and contribute to economies in steel detailing and fabrication. The entry canopy also deviates from the previous examples in its replacement of the catenary edge cable termination with a fixed termination at the tubular steel edge beam. The fixed edge contributes to the clear and simple geometry of the structure. It also provides direct water runoff from the roof into continuous gutters, a decisive advantage in an entry structure.

The dramatic potential of the inverted cone at the Kaleidoscope Shopping Center is enhanced by the eight open-mesh perimeter "sails."

The entry canopy also illustrates the sometimes visually evocative nature of fabric roofs, particularly those tented forms that have been likened to mountains, sails, or other forms. I was surprised, however, at the likeness apparently discovered by numerous passersby at the entrance canopy. While I was photographing the structure, a young man approached me. "It's National Semi's tribute to Madonna," he joked, in reference to the cone-shaped brassieres worn by the pop star in her Blonde Ambition show. He went on to say that the structure had drawn feminist ire for the suggestiveness of its shapes. While some architects consciously seek imagery in their fabric roof designs, the painted steel cones at National Semiconductor were selected as a simple and straightforward termination of the fabric form, with no thought of breasts or brassieres.

An interesting variation of the classic cone is the inverted cone, in which the fabric necks down from its perimeter to a low point at the center. There are several practical reasons for avoiding such configurations, including the tall walls or supporting columns required around the perimeter, the large obstruction of the center of the space incurred by the low point of the roof itself, and the necessity for internal drainage. More often than not, the configuration is chosen for dramatic effect.

My firm engineered an inverted cone for the Kaleidoscope Shopping Center in Mission Viejo, California (shown above). The structure is supported on eight tubular steel bents, each of which provides attachment for both the perimeter edge catenaries and the tension hoop at the central low point. As conceived by architect Altoon & Porter, the structure is decorative in function. The inverted cone, in combination with eight open woven perimeter "sails," has the airy beauty of a blossoming flower.

THE TENSIONED FABRIC ROOF

The arched form of the Milano Fair Ground roof is achieved with a suspended roof system rather than a conventional arch.

Source: Canobbio S.p.A.; used with permission.

Suspended Roofs

By suspending the fabric membrane from cables hung off one or more masts or other compression members, designers can create long-span roofs of almost any shape. Practically, the cost of such roofs is burdened by the length of the compression members that are required to reach far above the fabric. Aesthetically, the exposed skeletons of these roofs provide the opportunity for visual drama, while the suspended shapes themselves range from amorphous and unsatisfying to the elegantly sculpted. Massimo Majowiecki's design of the Milano Fair Ground roof (shown above) is a successful application in which the roof is supported by vertical suspension cables hung from four parallel catenary shaped cables that lift the ridge line of the fabric into an arched form. While the means by which the fabric is supported is entirely different, the membrane form is reminiscent of the Munich Ice Rink discussed in the section on arch systems.

Nearly 20 years after its construction, the Hajj Terminal in Jeddah remains both the world's largest tensioned fabric roof and one of the great achievements of the technology.

Source: Birdair, Inc.; used with permission.

While air-supported roofs and cable domes are generally the most economical systems for use on long-span fabric roofs, the two largest fabric structures built to date—the Hajj Terminal in Jeddah, Saudi Arabia, and the Millennium Dome in Greenwich, England—both employ suspended roofs. Completed in 1981, the Hajj Terminal (shown above) provides temporary shelter for travelers landing at the Jeddah airport who are awaiting bus transport for continuation of their pilgrimage to Mecca. The roof is vast in all ways: its physical scale, its visual impact, the complexity and geographic diversity of the team that designed and built it, and the scope of its technical achievements. As designed by architect SOM, it has a total area of 425,000 m^2 and comprises 210 cone-shaped modules, each 45 m square in plan, that are combined into 10 groupings that are each 3 modules wide by 7 modules long. The fabric forms themselves are conventional: each symmetrical cone reinforced by 32 radial cables, rising to a center ring of approximately 4-m diameter, with the fabric terminating in catenary cables at the perimeter of each 3-by-7 grouping.

It is the support of the peak of each module that is unconventional, however. Rather than place a vertical mast directly beneath each center ring as in a conventional cone, the architect elected to provide a mast at the corner of each module that extends 12 m above the top of the fabric, with a pair of suspension cables extending from the top of the mast down to the tops of each of the adjoining center rings. The system selected was not the most economical, as the cost of an additional mast directly supporting the peak of each module would have been less than the cost that was incurred by extending each of the corner masts to well above the top of the canopy and adding suspension and stabilizing cables (*Architectural Record* 1980).

England's Millennium Dome duplicates the form of a domed compression structure in a suspended fabric tensile structure.

Source: FTL Happold; used with permission.

[ABOVE] Exterior view.

[FACING PAGE] Interior view.

Because of its extraordinary scale, novel design, and reliance on relatively new analytical techniques (see Chapter 5), extraordinary measures were taken to ensure the safety of the roof, including verification that failure of any individual cable did not lead in dominolike fashion to the failure of additional elements. Intuition alone tells us that such checks are particularly important in suspended roofs, where the advance of corrosion in high-strength cable wires of small cross section or unreliability in an individual swaged fitting or other cable connection element should not be permitted to lead to instability in the overall roof form. As a design engineer at Geiger Berger, engineering consultant to roof contractor Owens Corning Fiberglas, I participated in numerous cycles of analysis in which the roof's overall stability in the face of failure of individual suspension, radial, and catenary cables was confirmed. These analyses included successful prediction of the forces in remaining cables and other elements when individual cables were released to simulate failure on a prototype constructed of two full-scale modules.

The 80,000-m² plan area of the Millennium Dome is about 20 percent of the Hajj Terminal area. While the Hajj Terminal is broken into a multiplicity of modules whose form and relatively small scale provide a reminder of nomadic tents, however, the single 320-m sweep of the Millennium Dome has a grandness befitting the once in a thousand year event that it commemorates. "A dome is the universal symbol of assembly," says Michael Davies of architect Richard Rogers Partnership (*Architecture* 1999). Working in partnership with structural engineer Buro Happold, Rogers has for the first time brought the British talent for elegantly expressive structures seen on a number of smaller roofs (Mound Stand at Lord's Cricket Ground, Schlumberger Research Centre, and others) to bear on a major fabric roofed assembly building. Previously, this field had been dominated almost entirely by American air-supported and cable dome designs, with a few similar Japanese creations.

By bringing the elegantly curved dome almost to grade at the perimeter, the Millennium Dome designers avoided the coarsely formed and detailed supporting structures that mar most of the air-supported and cable dome designs, and maintained a purity of structural form exceeded only by the United States Pavilion at Osaka (shown on page 44).

The Millennium Dome roof is constructed with 12 trussed steel towers from the tops of which suspension cables reach down to support 72 radial cable lines. As at the Hajj Terminal, stabilizing cables running from the underside of the membrane to points low on the mast are used to resist wind uplift forces on the membrane.

The structural system of the Millennium Dome is not likely to replace systems such as the cable dome for long-span applications, however. The long compression members (approximately 100 m) that

[ABOVE] The repeated ridge and valley modules of the enormous Denver Airport roof provide its classic sawtooth form.

[RIGHT] Those journeying through Denver pass through a basilica for the contemporary traveler.

Photographs copyright Robert Reck.

it shares with other suspended roofs limit its economy, and placement of the 12 masts approximately 60 m in from the edge of the building makes it an unfavorable choice for maintaining sightlines for athletic events or similar uses (ENR 1998).

Ridge and Valley Systems

Scalloped forms with fabric restrained by alternating ridge and valley cables provide the opportunity for a repeated modular form, where architecturally appropriate. Designed by architect Fentress Bradburn and engineer Horst Berger, the Denver Airport roof (shown above) provides the most prominent example of a classic multi-bay ridge and valley roof. Vertical masts near the outside edges of the roof provide a termination for the peaks of the roof modules and the ridge cables crossing the roof. Valley cables span across midway between adjoining pairs of ridges to provide the characteristic scalloped form of a ridge and valley system. From the inside, the undu-

The long-span ridge and valley system of the San Diego Convention Center is achieved by the use of cable-suspended masts.

[ABOVE] Exterior view.

Photograph copyright Robert Reck.

[LEFT] Interior view.

Source: Ronald Schaeffer; used with permission.

lating form and the glazed sidewalls are reminiscent of an enormous basilica. From the outside, the peaked forms mirror the shapes of the (sometimes) snow-capped mountains beyond with some success, as desired by the architect. The architect varied the height of the peaks at the two ends and at the two main crossing points of the terminal in reflection of the variation of function at these locations. These variations in height may detract from the orderliness of the design when it is viewed from oblique angles, however, and the economy and generally satisfying appearance of the roof might both have been improved by maintaining regularity in height throughout the 270-m length of the roof.

[ABOVE, ABOVE RIGHT] The bravura of the Jameirah Beach Hotel canopies derives from the open eyes created by their ridge cables and from valley cable terminations that "float" between sloping masts.

Source: Paul Roberts; used with permission.

[ABOVE] The structural system for an amphitheater roof such as that at Chang Sha, China, must integrate the sometimes conflicting geometric demands of structure seating, circulation, sightlines, and acoustics.

Source: Beijing N&L Fabric Techology; used with permission.

Economical spans of roof and valley systems are limited by the flatness and lack of stiffness of the roof near midspan. At the Denver Airport, for example, upward and downward deflections of approximately 750 mm were calculated under design loads (DePaola 1994). At the San Diego Convention Center (shown on the previous page), created by Berger with architect Arthur Erickson, the bottoms of the masts were supported on overhead cables in order to increase the clear span while maintaining an elegantly low profile.

While "flying" masts were required to maintain clear spans in San Diego, the aisles between seating sections provided appropriate mast base locations for the ridge and valley roof that my firm engineered for the amphitheater in Chang Sha, China (shown to the left). Architect Beijing N&L Fabric Technology selected initial mast heights as required to provide a steady rise in mast peak elevations moving from the stagehouse to the back of the seating, in order to mirror the rise in grade from front to back. Spans between masts vary substantially in reflection of the width of the seating area, however. We made adjustments in mast height when engineering the structure to maintain a reasonable vertical offset between ridge and valley cables near midspan and to provide enough stiffness to keep the fabric stable in this area under wind load.

Broad variations in roof form can be attained while adhering to the overall theme of alternating ridges and valleys. The Palm Court Canopies designed by architect W.S. Atkins for Dubai's Jameirah Beach Hotel (shown above), terminate at the roof high point, so that their form is essentially half of a ridge and valley system such as that used at Denver. In addition, Jameirah Beach replaces the conventional ridge cable with a pair of catenaries to create "cat eye" openings between adjacent fabric modules.

Evolutions in the materials and design methodologies of tensioned fabric structures now give them a place in high grade architecture produced by the best designers and specialty contractors. Their drama and visual appeal find frequent use as dramatic elements in larger structures. The open petals of this inverted tent provide a focal point for a suburban shopping center.

KALEIDOSCOPE SHOPPING CENTER

Designers: Altoon + Porter,
 Huntington Design Associates
Builder: Fabritec, with Eide Industries

[ABOVE] Overall
[LEFT] Beneath

THE TENSIONED FABRIC ROOF **PLATE 1**

The simplest uses of fabric structures include minimally curved entry or shade structures. Well articulated and carefully detailed supporting members and connections distinguish these structures from conventional awnings, while the effects of natural and artificial light on translucent and reflective planes of white fabric provide ever-changing delight.

BURNSIDE RESIDENCE

Designers: Archaeo Architects,
  Huntington Design Associates
Builder: Rader Awning & Upholstering

[ABOVE, LEFT] Living Room Canopy
  Archaeo Architects; used with permission
[ABOVE, RIGHT] Walkway Canopy
  Photograph copyright Robert Reck; used with permission
[RIGHT] Courtyard Canopy
  Photograph copyright Robert Reck; used with permission

PLATE 2 THE TENSIONED FABRIC ROOF

HOLLYWOOD & HIGHLAND

Designers: The Ehrenkrantz Group, Altoon + Porter, Huntington Design Associates
Builder: Sullivan & Brampton

[LEFT] Reflected Light
[BELOW] Escalator Canopy at Dusk
Don Douglas; used with permission

THE TENSIONED FABRIC ROOF PLATE 3

Other entry canopies explore variations on the traditional curvilinear tented form. They may be formed by guyed struts and overlapped in asymmetrical fashion, or generated by steel arms that curve to mirror the form of interlinked inverted cones.

WHITE LOTUS RESTAURANT

Designer: Huntington Design Associates
Builder: Eide Industries

[ABOVE] Overall
[RIGHT] Detail
  Eide Industries, Inc.; used with permission

PLATE 4 THE TENSIONED FABRIC ROOF

PARKWEST MEDICAL OFFICE BUILDING

Designer: Huntington Design
 Associates
Builder: Eide Industries

[ABOVE] Overall
[LEFT] Strut End Detail
 Eide Industries, Inc.; used with permission

THE TENSIONED FABRIC ROOF **PLATE 5**

The form of the Folkestone Terminal for the English Channel Chunnel Train is exotic, with its cable-suspended mast providing a column-free open space below. The traditional form of a simple train shed is adapted for the Bay-pointe light rail station, with a break in form provided by a vaulted crossing.

FOLKESTONE CHUNNEL TERMINAL

Designer: Building Design Partnership
Builder: Birdair

[BELOW] Overall
[RIGHT] Interior Detail
QA Photos Ltd.; used with permission

PLATE 6 THE TENSIONED FABRIC ROOF

BAYPOINTE STATION

Designers: SBA Architects,
 Huntington Design Associates
Builder: Birdair

[ABOVE] Overall
[LEFT] Interior
[BELOW] Detail

THE TENSIONED FABRIC ROOF **PLATE 7**

Just as the wings of early aircraft were formed by cloth, the forms and spirit of tensioned fabric structures seem particularly attuned to terminals for air travel. Each structure, though, must adapt to its local environment. The Hajj Terminal in Jeddah, Saudi Arabia, is open sided and top vented to release the desert heat, while the form and load design of the insulated Denver Airport reflects its snowy and mountainous surroundings.

HAJJ TERMINAL

Designers: Skidmore Owings & Merrill, Geiger Berger Associates
Builder: Owens-Corning Fiberglas with Birdair

[ABOVE] Overall
[RIGHT] Interior
Birdair, Inc.; used with permission

PLATE 8 THE TENSIONED FABRIC ROOF

DENVER AIRPORT

Designers: C.W. Fentress,
J.W. Bradburn & Associates,
Horst Berger with Severud
Associates
Builder: Birdair

[ABOVE] Overall
[LEFT] Detail
[BELOW] Interior
  Photographs copyright Robert Reck

THE TENSIONED FABRIC ROOF **PLATE 9**

Hospitality applications also adapt to divergent environments. The Jameirah Beach Hotel canopies provide ventilated shading for the desert heat of Dubai. The high desert of Taos, New Mexico, though, has enormous seasonal temperature variation. The fabric roof of the interior oasis at the El Monte Resort is designed to support winter snow loads, while operable mast-top ventilation with electric fans and shading side panels provide protection from the summer heat.

JAMEIRAH BEACH HOTEL

Designers: W.S. Atkins, Huntington Design Associates
Builder: Birdair

[RIGHT] Shade Canopy
[BELOW LEFT] Shade Canopy Detail
[BELOW RIGHT] Shade Canopy
  Paul Roberts; used with permission

PLATE 10 THE TENSIONED FABRIC ROOF

EL MONTE RESORT

Designers: Dharma Properties,
  Nims, Calvani & Associates,
  Huntington Design Associates
Builder: Eide Industries

[ABOVE] Overall
[LEFT] Interior
  Eide Industries, Inc.; used with permission

The inherent drama of fabric structure forms gives them natural application in assembly or entertainment venues such as amphitheaters. Peaked and curving forms bring words to the mouths of visitors approaching from distant streets, parking lots, or freeways. "There it is!"

NATIONAL SEMICONDUCTOR AMPHITHEATER

Designers: Bellagio Associates,
   Huntington Design Associates
Builder: Sullivan & Brampton

[RIGHT] Interior
   Don Douglas; used with permission
[BELOW] Overall
   Bernard André; used with permission

PLATE 12   THE TENSIONED FABRIC ROOF

WEBER POINT EVENTS CENTER

Designer: Huntington Design
  Associates
Builder: Sullivan & Brampton

[ABOVE] Overall
[LEFT] Detail
  Don Douglas; used with permission

Convention and exhibition centers take advantage of the long span capability of fabric construction, though their forms vary widely. San Diego's waterfront canopies are frankly sail-like, while Genoa's Grande Bigo responds to its harbor site with steel rigging reminiscent of ship loading derricks.

SAN DIEGO CONVENTION CENTER

Designers: Arthur Erickson, Horst Berger Partners
Builder: Birdair

[ABOVE LEFT] Overall
  Photograph copyright Robert Reck
[ABOVE RIGHT] Interior
  Photograph copyright Robert Reck
[RIGHT] Detail
  Ronald Shaeffer; used with permission

PLATE 14 THE TENSIONED FABRIC ROOF

EL GRANDE BIGO

Designers: Renzo Piano, Ove Arup
Builder: Canobbio

[ABOVE] Overall
[LEFT] Detail
   Canobbio S.p.A.; used with permission

THE TENSIONED FABRIC ROOF **PLATE 15**

Arched forms typically provide softer and more sheltering forms than peaked tents. Suspended arch forms, though, can have a special drama: the Munich Ice Rink hangs the membrane delicately from a graceful steel arch, while the downward curving cables that support the Milano Fairgrounds roof are an inversion of the arched form of the membrane itself.

MUNICH ICE RINK

Designers: Ackermann & Partner,
    Schlaich Bergerman & Partners
Builder: Maurer Söhne, Koit-Werk

[ABOVE] Arched Exterior
[RIGHT] Interior Cable Net
   Jörg Schlaich; used with permission

PLATE 16 THE TENSIONED FABRIC ROOF

MILANO FAIRGROUNDS

Designers: Massimo Majowiecki,
 Casalecchio di Reno
Builder: Canobbio

[ABOVE] Overall
[LEFT] From Grade

THE TENSIONED FABRIC ROOF PLATE 17

Domed form allows a stone or concrete shell to span long distances in pure compression—structural behavior that is antithetical to that of a tensioned fabric roof. The Pontiac Silverdome duplicates dome form through internal air pressurization that creates the pure tension dome of the low-profile air-supported roof, while the Millennium Dome does so by suspending the fabric from tall masts.

PONTIAC SILVERDOME

Designers: O'Dell Hewlett & Luckenbach, Geiger Berger Associates
Builder: Birdair

[RIGHT] Overall
[BELOW] Interior
  Birdair, Inc.; used with permission

PLATE 18 THE TENSIONED FABRIC ROOF

MILLENNIUM DOME

Designers: Richard Rogers Partnership, Buro Happold
Builder: Birdair

[ABOVE] Overall
QA Photos; used with permission
[LEFT] Interior
FTL Happold; used with permission

THE TENSIONED FABRIC ROOF   **PLATE 19**

Dome form is maintained (and a "true" dome's compressive load-carrying behavior absent) in the cable domes that have supplanted low-profile air-supported roofs. The Tropicana Dome employs the original and straightforward Geiger cable dome system, while the Georgia Dome employs the Levy "hypar-tensegrity" system—at once more complex structurally and more poetic visually.

TROPICANA DOME

Designers: Hellmuth, Obata + Kassabaum, Geiger Engineers
Builder: Birdair

[ABOVE] Overall
[RIGHT] Interior

Photographs copyright Robert Reck

PLATE 20 THE TENSIONED FABRIC ROOF

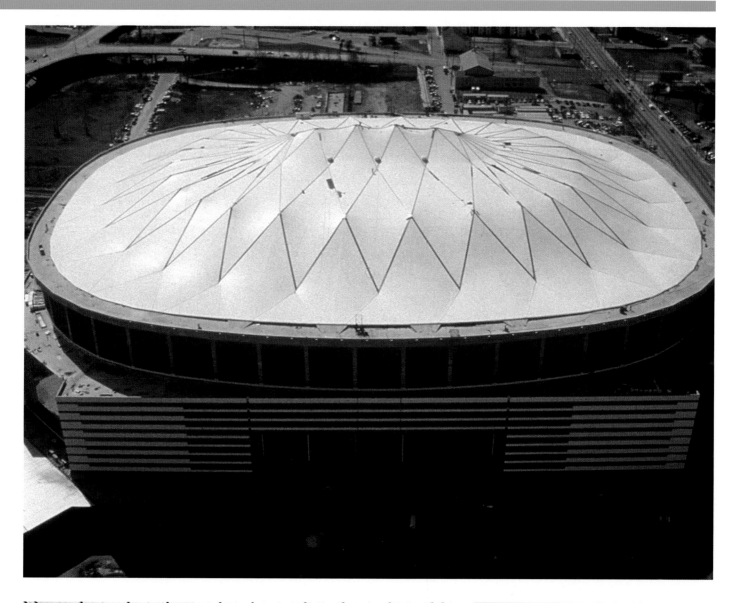

GEORGIA DOME

Designers: Heery Architects &
 Engineers, Weidlinger Associates
Builder: Birdair

[ABOVE] Overall
[LEFT] Interior Cable Dome Structure
Photographs copyright Robert Reck

THE TENSIONED FABRIC ROOF PLATE 21

Stadiums in which only the seating area is covered require different design solutions than "domes." In King Fahd stadium, the huge oculus that opens the playing field to the sky is restrained only by a cable tension ring. When a grandstand is linear, though, the free edge of the canopy requires rigid support, such as Avery Aquatic Center's cantilevered pipe arches.

KING FAHD STADIUM

Designers: Ian Fraser, John Roberts & Partners, Schlaich Bergerman & Partners, Horst Berger Partners
Builder: Birdair

[ABOVE] Overall
[RIGHT] Interior with Tension Ring
  Birdair, Inc.; used with permission

PLATE 22 THE TENSIONED FABRIC ROOF

AVERY AQUATIC CENTER

Designers: ELS/Elbasani & Logan Architects, Huntington Design Associates
Builders: Sullivan & Brampton

[LEFT] Overall
[BELOW LEFT] Interior
[ABOVE] Tensioning Rod Detail
  Don Douglas; used with permission

THE TENSIONED FABRIC ROOF PLATE 23

The steep upward sweep of a single-peaked cone and the ethereal interior light provided by architectural fabrics have natural application in a church. Fabric structures find their place in the fulfillment of man's highest aspirations.

GOOD SHEPHERD CHURCH

Designer: Gene Zellmer
Builder: Birdair

[ABOVE] Exterior
[RIGHT] Interior

Cantilever Canopies

Cantilevers encompass the family of structures in which one or more edges of the roof are supported, not from below, but by struts or arches cantilevered off a point on the interior or far edge of the structure. They are often appropriate where the structure's use demands unobstructed sightlines to points beyond its perimeter. They assume a variety of forms, with the struts cantilevering to one or both sides of the support or even radiating out from a central mast like spokes. Radial cantilevers were employed on umbrellas that my office engineered for the White Lotus Restaurant (Plate 4) and the Jameirah Beach Hotel (shown above; see also the photograph on page 132).

King Fahd stadium in Riyadh, Saudi Arabia (shown on the next page), makes use of an interesting variation of the conventional cantilever conceived by Horst Berger to create a structure with an overall width of 270 m. Support masts are placed in a circular configuration, and a circular tension ring is provided at the inside edge of the cantilevered fabric. The inward "pull" of the ring at each angle change in the cable replaces the "push" provided by the horizontal struts in a conventional cantilever. The large oculus in the center leaves the playing field in direct sunlight, providing daylight sufficient for healthy grass growth while completely covering the grandstands. Much like a ridge and valley system, of which this might be considered a variation in radial form, the ultimate spanning capabil-

On the Jameirah Beach Hotel Sundeck Stair canopy, the vertical tie-down cables used at White Lotus are replaced by cables that stretch from the ends of the struts down to a common workpoint lower on the mast (see also page 132). This alternative strut restraint has the advantages of limiting foundation work and obstructions at grade associated with the vertical cables.

Source: Paul Roberts; used with permission.

King Fahd stadium takes the conventional linear arrangement of grandstand cantilevers and turns it back onto itself in a circular form.

Source: Birdair, Inc.; used with permission.

[ABOVE] Exterior view.

[RIGHT] Interior view.

ity of these structures is limited by the flatness of the fabric surface in the vicinity of the tension ring.

Grandstand covers commonly make use of cantilevered systems, with support provided only at the top of the grandstand in order to provide unobstructed sightlines to the field below. The paired arches of the grandstand cover for the Avery Aquatic Center of Stanford University, on whose design we collaborated with ELS architects, are configured to provide a short cantilever over the upper walkway and a longer cantilever over the grandstand seating (shown at the top of the next page). Front and rear arch cantilevers are shop-welded to sloped supporting struts that are in turn field-bolted to the tops of vertical pipe columns. The columns are configured in pairs, with a steel pipe anchored to either side of the concrete girders supporting the grandstand seats.

[ABOVE] The cantilevered canopy system used at Stanford University's Avery Aquatic Center provides shading for both the walkway above the supporting columns and the grandstands below. The light weight of the fabric system was ideally suited for retrofit on the existing concrete structure.

Source: Robin Huntington; used with permission.

[LEFT] The radiating cantilevered struts used at the Parkwest Medical Office Building mirror the curved façade of the building behind. The four inverted tents drain into conical steel pans that collect the water into drain leaders behind each column.

Source: Eide Industries; used with permission.

In our design for the entrance canopy of the Parkwest Medical Office Building in Knoxville, Tennessee, four inverted cones are supported on arches that cantilever radially, like tree branches, from center posts (shown above). The arches are braced against upward and downward loads by pipe struts spanning from the mast peaks to points near the ends of the arches. Because the struts are above the fabric, erection required that the membrane be placed atop the arches before bolting the struts into position above and tensioning the membrane.

The parallel arches of the Hanover Park, Illinois, tennis facility [ABOVE RIGHT] rely on interior cross-bracing for stability [ABOVE LEFT].

Source: Birdair, Inc.; used with permission.

[RIGHT] The crossed arches used at the Pizzitola Athletic Facility eliminate the need for lateral bracing.

Source: Birdair, Inc.; used with permission.

Arch Systems

Arches are the primary generators of saddle fabric-roof forms. Their use became ubiquitous in the 1980s for 10 to 20 m span atrium covers in strip shopping malls. Typically, such applications employ a repeated arch spacing that corresponds with the structural module for the adjoining buildings. Steel pipe arches are typically employed on structures of moderate span, although trussed pipes, glu-laminated timber, and even concrete also are used. Arches sometimes are placed perpendicular to the axis of the atrium, a configuration that has the drawback of requiring that the arches either be braced in some fashion or rigidly fixed at their ends to prevent them from toppling laterally (shown above). More often, pairs of adjoining arches are rotated in plan and crossed at their peaks in an "x" pattern to laterally stabilize each other (also shown above).

Canted arches provide an alternative to crossed arches. Their design is a refinement of the conventional parallel arch in which the

Canted arches, such as at Buena Ventura Shopping Center, brace each other while providing a lively variability of curved roof forms.

[ABOVE] Exterior view.

[LEFT] Interior view.

peaks of adjacent arches are displaced toward each other until they meet at a tangent and are connected to each other in order to provide lateral stability. This approach was used by architect Charles Kober at the Buena Ventura Shopping Center in Ventura, California (shown above), a structure on which my firm provided consulting services to roof contractor Birdair. The photo illustrates the liveliness that these forms typically have in comparison to crossed-arch and parallel-arch structures. Small valley cables were added perpendicular to the main axis of the structure and the paired arches in order to draw the fabric farther downward between the arches and provide greater articulation of the form.

Arch-supported roofs are less efficient for long spans than either the air-supported or cable dome systems discussed below, because they rely on compression members that span the full width of the roof. Their forms are generally more restrained than those of cones, although single long-span arches have occasionally been used with great dramatic effect. At Lindsay Park Sports Centre in Calgary,

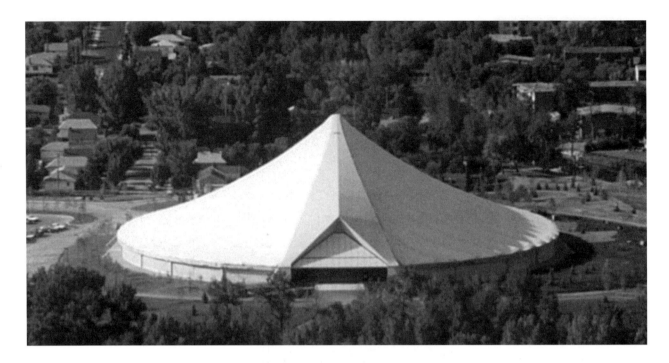

[ABOVE] Lindsay Park's muscular arched roof attains a sculptural quality at the expense of visual lightness.

Source: Birdair, Inc.; used with permission.

[RIGHT] Lindsay Park Sports Centre.

[BELOW RIGHT] Cross section through roof construction.

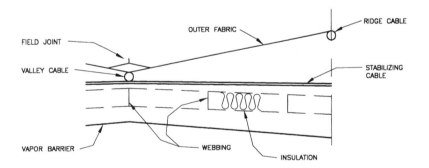

Alberta, designed by architect Chandler Kennedy and Geiger Berger engineers (shown above), arch stability is achieved by the broad base of the single 120 m-span arch that effectively provides a moment-connected base.

Lindsay Park's sculptural approach is carried through in the design of the fabric membrane itself. The roof has alternating ridge and valley cables perpendicular to the arch, giving it a scalloped form. Unlike a conventional ridge and valley system in which cables

Both the carefully articulated exterior arch [TOP AND ABOVE LEFT] and the interior cable net [CONSTRUCTION PHASE, ABOVE RIGHT AND RIGHT] of the Munich Ice Rink contribute to its lean, skeletal quality.

Source: Jörg Schlaich; used with permission.

curve alternately upward and downward, however, both sets of cables are cupped upward, and anticlastic stability is provided by the convex curved stabilizing cables parallel to the arch. While the supporting arch is at once elegant and muscular in form, the roof as a whole has a visual mass that belies its nature as a lightweight fabric membrane. Seen from a distance, the roof could as easily be imagined to be concrete or some other rigid material as it is fabric.

Lindsay Park stands in sharp contrast to the structural minimalism of the similarly sized Munich Ice Rink (shown on this page), whose roof was designed by Jörg Schlaich with architect Ackermann and Partner. The minimally curved fabric membrane of Munich is shaped by an internal cable net, which is in turn supported from the

The Institute for Lightweight Structures at the University of Stuttgart was constructed as a prototype for the German Pavilion at Expo '67 in Montreal. A cable loop, in-filled with a skylight, gathers the accumulated tensions at the top of the cable net and distributes them into the mast.

Source: Larry Medlin; used with permission.

single exposed steel three-chord arch by vertical suspension cables. Munich's most daring statement lies in the means used to attach the fabric to the arch, however. The arch is neither hidden below the fabric nor in the plane of the fabric, as at Lindsay Park, but stands nearly 2 meters above it. While the vertical suspenders from the arch support the cable net, the cable net in turn provides lateral stabilization to the arch, due to the equilibrating effect provided by the lateral component of force in the suspenders as the arch begins to sway laterally. Cable loops between the suspenders collect the cable net tensions while providing openings for glazed areas underneath the arch.

The design is analogous to the cable loop used by earlier German designers to connect cable nets or fabric to mast (shown above). The result at Munich is delicate and structurally expressive, and the tension-to-compression load transfer is delineated so exquisitely that it rivals the best of the mast-supported structures for visual drama.

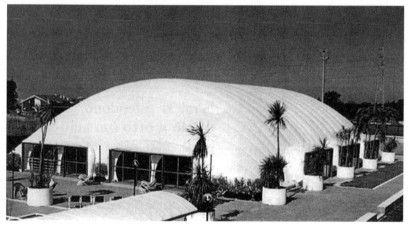

[ABOVE LEFT] The early Radomes established the ability of fabric designers to create reliable roofs of simple form without benefit of complex analysis procedures.

Source: CHEMFAB Corporation; used with permission.

[LEFT] The swimming pool cover of Rome's Realmonte Sporting Club is a successful adaptation of the original air-supported Radome technology.

Source: Canobbio S.p.A.; used with permission.

Air-Supported Roofs

Air-supported fabric roofs have been in use since the construction of Walter Bird's first 15-m-span "Radome" (shown above, top) in 1946 (Shaeffer 1996). Reliant as they were on simplified methods of hand analysis, the Radomes employed tall, simple spherical shapes, relatively high air pressures, and the experience gained on repetitive designs for their reliability.

Following their perfection of the spherical air-supported roof, Bird and his collaborators at Birdair, as well as others, expanded Radome technology to include the semi-cylindrical shapes now sometimes called "tennis bubbles" (shown above). The newer forms shared the high profile of the Radomes and their relative ease of analysis, while providing an easily erected and economical means of enclosing large spaces. The roofs remain competitive with prefabricated metal buildings for storage and recreational facilities with fairly long spans.

[ABOVE] The United States Pavilion in Osaka, Japan, whose design was selected for its great economy, had profound impact on the next 15 years of long-span construction.

Source: Geiger Engineers; used with permission.

[RIGHT] Like the top surface on an airplane wing, a low-profile air-supported roof like the United States Pavilion in Osaka experiences upward forces across the entire roof surface.

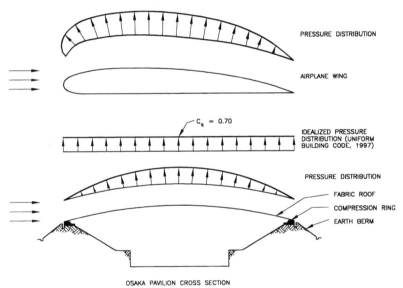

With this background, the low-profile air-supported design developed by engineer David Geiger for the United States Pavilion at Expo '70 in Osaka, Japan (shown above), represented a true revolution in design. The critical change made by Geiger was to give the roof a low and minimally sloped profile, with the shortest span of the oval shape nearly 13 times the roof's rise. By doing so, he was able to take advantage of a key aerodynamic principle: roofs of low slope are subject only to upward "lift" forces due to wind, much like the lift on the top surface of an airplane wing (see the drawing above) (Lodewijk 1967; Ishii 1995). While inward wind pressure on the steeply sloped walls of a Radome (see the drawing on next page) must be resisted by the high internal air pressure, the wind on a low-profile roof may be resisted by an increase in tension in the roof membrane alone.

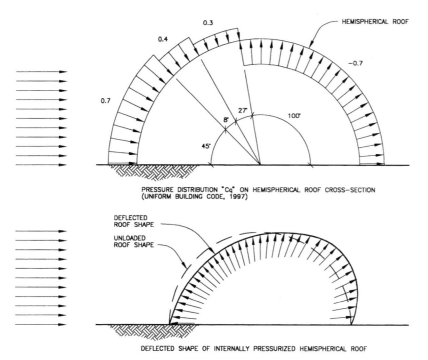

PRESSURE DISTRIBUTION "Cq" ON HEMISPHERICAL ROOF CROSS-SECTION (UNIFORM BUILDING CODE, 1997)

DEFLECTED SHAPE OF INTERNALLY PRESSURIZED HEMISPHERICAL ROOF

On a high-profile roof such as a Radome or tennis bubble, high internal pressurization is required to resist large distortions or instability under wind load.

Just as a flat profile in a suspension bridge cable or a prestressed concrete tendon results in higher cable forces, the advantages of the low-profile roof are achieved at the cost of high-tension stresses in the fabric membrane. Generally, this is overcome by the addition of cabling laid over the fabric to reinforce the membrane. The fabric effectively spans from cable to cable rather than across the entire roof and is patterned to dimple upward between cables to provide small curvature radii. At Osaka, as on his later designs, Geiger laid the cabling diagonally over the roof in a pattern of "skewed symmetry" that minimizes the bending moment in the perimeter ring beam and places it in nearly pure compression, by which it can resist loads most efficiently.

The forces on a low-profile air-supported roof are directly analogous to (but opposite) those in a conventional domed roof. The air roof is subjected to upward loads that place the membrane in tension resisted by a perimeter compression ring, while the dome is subjected to downward dead and live loads that place the dome in compression resisted by a perimeter tension ring. The air roof is ideally adapted to the structural properties of fabric, which has high tensile strength but negligible compressive resistance. Domes, by contrast, are characteristically constructed of concrete, a material of high compressive strength and limited tensile strength.

Following the completion of Osaka and the development of long-lived polytetrafluoroethylene (PTFE) coated fiberglass fabrics (see Chapter 4), Geiger further developed the low-profile air-supported roof to use on stadiums with spans of more than 200 m. The Silverdome of Pontiac, Michigan (shown on the next page) was the first of the major air-supported stadium roofs engineered by the Geiger

THE TENSIONED FABRIC ROOF

The Pontiac Silverdome was the first of the stadium-sized air-supported roofs.

Source: Birdair, Inc.; used with permission.

[ABOVE] Exterior view.

[RIGHT] Interior view.

Berger office and constructed by Birdair. In addition to its advance in scale, Pontiac demonstrated clearly that the air-supported fabric roof provided unmatched economy in stadium-sized enclosures.

The long-span air-supported roofs in Pontiac and other cold locales are designed so that air pressure can be increased to counteract snow buildup. They generally have fans to pump heated air into the gap between the structural membrane and a liner membrane below, in order to melt any snow or ice buildup before loads become excessive. In addition, the roofs are designed so that the cabling generally limits failure to a single panel of fabric and so that the fabric drapes well above the grandstands should a deflation occur.

The design safeguards used on the air-supported roofs have given them a perfect life-safety record, but they have not prevented numerous and sometimes costly full or partial deflations. Such occurrences at Pontiac raised the first doubts about the technology. The problems were, unfortunately, duplicated in other major stadiums, and generally resulted from a combination of severe storms

Air-inflated roofs match the economy of their air-supported cousins while providing the opportunity for open-sided applications, such as those required for greenhousing.

with either failure of the mechanical system or failure to activate it until snow buildup had already occurred. Several of the deflations have been costly, due to repair cost or event cancellation, and they have resulted in too many column inches of press coverage that has soured the public perception of low-profile air-supported roofs. No major new air-supported roofs have been constructed in North America since the completion of Vancouver's B.C. Place Stadium in 1983.

Air-supported roof development continued in Japan, however, with the completion of the Tokyo Dome in 1988. This structure, designed using the technology developed by Geiger, has not, to date, suffered deflations or other major problems. This fact is attributable in part to the support provided by Takenaka, which keeps a permanent office in the stadium and maintains computer-activated control of the heating and pressurization systems that protect the roof from deflation. That arrangement reflects larger differences in the nature of the construction industries in the United States and Japan. "In Japan," says Taiyo Kogyo's Moto Nohmura, "the general contractor has total control, and air roofs can work" (Nohmura 1992, personal communication). In spite of the Tokyo Dome's excellent record and management, however, events are sometimes cancelled as a precautionary measure in the event of inclement weather, and the Japanese have made no further ventures into long-span air-supported roof construction (Nohmura 1992, personal communication).

Among those disappointed by the halt in the development of long-span air-supported roofs is Walter Bird, creator of the original Radome. In lamenting the widespread fear of constructing new air roofs, he recalls his early training as an aeronautical engineer. "An airplane has to perform right 100 percent of the time, too, but that doesn't keep us from flying planes. Until people recognize air roofs as dynamic structures that must be properly designed *and* maintained, there will be problems" (Bird 1999, personal communication).

The ancient amphitheater at Nimes was revitalized by the addition of an air-inflated membrane roof.

Source: Jörge Schlaich; used with permission.

Air-Inflated Lenses

By encapsulating the pressurized air within a fabric lens at the roof level of the enclosed space, air-inflated structures avoid the need for an air-sealed building enclosure that limits the application of air-supported roofs. Because building occupants do not breach the air enclosure while entering and exiting the building, the energy required to keep the roof aloft is much less than for an air-supported roof.

In inflated roofs that my firm has engineered for greenhouse applications, fabric side panels beneath the lens can be rolled up to provide an open-sided configuration when weather permits (see photograph on page 47). These structures are built with curvature in the membrane only in the transverse section and are constructed in a series of bays of approximately 3-m width, each banded by heavy polyester webbing supported on a pipe steel frame. The uniaxial curvature results in membranes having minimal patterning and maximum economy, while providing the opportunity to extend the struc-

48   THE TENSIONED FABRIC ROOF

The amphitheater at Nimes.

Source: Jörg Schlaich; used with permission.

Steel outriggers lift the membrane from the arena floor [ABOVE] before it is inflated to its final lens form [LEFT].

ture longitudinally by adding bays, in the manner of a prefabricated metal building. Like metal buildings, their utilitarian appearance is best suited for agricultural or industrial applications, and the designer must beware that sufficient internal pressurization or other safeguards are taken to ensure that the structure does not ripple like a flag in the uncurved longitudinal direction in high winds.

An unusual air-inflated roof was constructed over the 2,000-year-old Roman amphitheater at Nimes, France (shown above and on the previous page). There were unique restrictions on the design, including the need to minimize load on the existing structure, and civic requirements that the roof not be visible from outside the arena and that it be dismantled each spring. The air-inflated roof concept developed by engineer Jörg Schlaich provides an inherent solution to the requirements for light weight and low profile, while his "self-erecting" design provides a clever solution to the different problems posed by seasonal erection and dismantling.

THE TENSIONED FABRIC ROOF   49

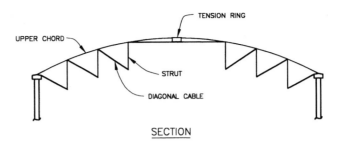

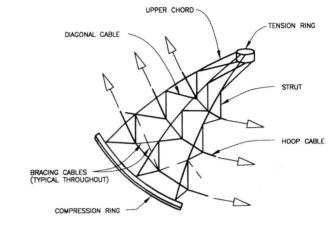

[ABOVE RIGHT] In section, the Geiger cable dome appears unstable because of the lack of a bottom tension chord in the plan of the truss.

[RIGHT] The three-dimensional view reveals the cable dome behavior: the stabilizing force at the bottom of each strut is provided by the angle change in the horizontal hoop cable. With the exception of the compression ring and struts, all members are in tension.

Cable Domes

Cable domes are a family of roof in which fabric (or other sheathing) is supported by a three-dimensional trusswork of vertical compression members and sloping tension members (cables) restrained by a perimeter compression ring. The simplest of cable domes is the king post structure (see the drawing on page 22). In general, though, the term "cable dome" is used to describe the more complex cable and strut systems used on stadiums and other long-span applications.

The first of the structures generally called a cable dome was created by engineer David Geiger as an alternative to his own air-supported systems. Because a fabric tear does not result in potential deflation or other instability, as it does in both air-supported and air-inflated structures, these roofs have demonstrated reliable performance on stadium-sized roofs. Part of their efficiency lies in the use of three-dimensional truss systems in which compression members are limited to short vertical struts.

An element of the cable dome's fascination lies in the fact that, when viewed in section, the cable truss appears unstable due to the lack of a bottom chord (see the drawings above). The tension in the diagonal cables exerts an outward pull at the base of each vertical post, which is equilibrated by the inward component of force exerted by a key structural element not apparent in a first glance at the section: the circumferential hoop cables that pull inward where they change angle at their attachment to the base of each post.

The cable dome roof of Florida's Tropicana Dome represents the structural system that has largely succeeded the air-supported roof for stadium applications.

Photographs copyright Robert Reck.

[ABOVE] Exterior view.

[LEFT] Interior view.

A Geiger cable roof is characterized by radiating cable trusses with triangular fabric panels between each pair of adjoining trusses (shown above). The steel elements are erected by tensioning, in turn, the diagonal cable attached to each of the hoops, beginning at the perimeter compression ring and working to the center. Tensioning of the diagonal cables introduces the appropriate prestress into each of the steel elements: tension in the hoop and ridge cables, compression into the posts. With the cable truss in equilibrium, the triangular fabric panels are installed between each of the ridge cables and tensioned in turn by stressing valley cables lain across the centers of the panels and extending from the perimeter compression ring to the center tension ring.

Structural engineer Matthys Levy of New York's Weidlinger Associates built upon the knowledge he gained as peer reviewer for Geiger's St. Petersburg, Florida, Tropicana Dome to develop the "hypar-tensegrity" system used on Atlanta's Georgia Dome (shown on the next page). Both the name and the basic structural system are

The hypar-tensegrity roof at the Georgia Dome is supported on a flower-like tracery of cables.

Photographs copyright Robert Reck.

[ABOVE] Exterior view.

[RIGHT] Interior view.

somewhat more complex than that of a Geiger cable dome. Specifically, the ridge and diagonal cables of the tensegrity roof do not radiate directly outward from a single center node (the tension ring), but step diagonally from hoop to hoop to form diamond-shaped panels into which the fabric is fitted following erection of the steel and its tensioning with the use of temporary cabling. In section, the tensegrity dome appears similar to the cable dome (see the drawings on the next page); but in plan, the change in cable connectivity results in a tracery of cables that blossoms outward from its center like a flower. The diamond-shaped fabric infill panels are smaller than the triangular panels of a Geiger cable dome, and they must be stretched into place between their clamped edges at the ridge cables. Erection and tensioning of the fabric is somewhat more complex, but Levy believes that the ability for redistribution of loads provided by the

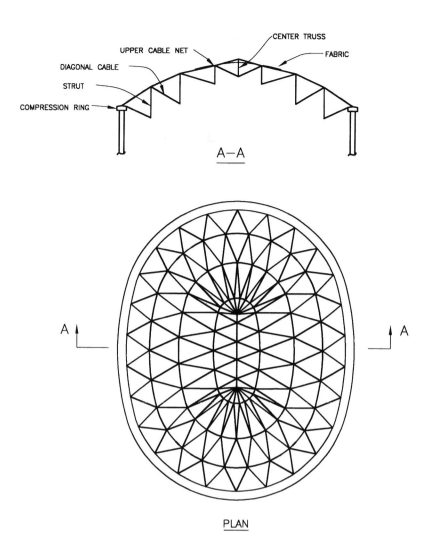

[ABOVE LEFT] In section, the hypar-tensegrity dome is similar to a Geiger cable dome.

[LEFT] The plan reveals a faceted pattern of upper cables that stabilizes and distributes loads around the dome in a manner that eliminates the bracing cables of the Geiger cable dome.

diagonally laid cables provides an improvement in the redundancy and consequent reliability of the roof.

In practice, both systems have offered good performance with no radical differences in economy. The Levy and Geiger camps have been unusually fervent in arguing the merits of their respective systems in both the technical literature and to all others with an interest. While structural engineers are typically noted for restraint in arguing the superiority of their own work, the verbal sparring has precedent. Dating back to the patent truss inventors of the nineteenth century and before, engineers at the most creative level of the profession have been fueled by passionate belief and economic interest to work hard in advancing the merits of their ideas.

Cable Nets

Cable nets do not represent a fabric-roof structural system so much as a means of strengthening and stiffening the fabric membrane that is applicable to nearly any of the systems discussed earlier in this chapter. They were originally developed by Frei Otto and his German associates for use with membranes of cotton and other lower strength

The cable net grid [ABOVE] of the Munich Olympic Stadium was infilled with clear acrylic panels following its erection [ABOVE RIGHT].

Source of photograph above: Jörg Schlaich; used with permission.

fabric, or where alternative roofing materials such as acrylic or wood cladding required direct support. Cable nets are often assembled in the shop, then rolled for shipment to the site. Most are constructed with the cables spaced to allow workers to stand on and walk across the net like a three-dimensional ladder.

The largest and best known of the German cable nets is that constructed for the 1972 Olympic stadium in Munich (shown above). The roof was designed by a team that included architects Günter Behnish and Frei Otto in collaboration with structural engineer Leonhardt und Andra and others, and that used some of the earliest computer software developed in Germany for analysis of membrane roofs. This software also permitted accurate determination of cable lengths and found the geometry of each of the clear acrylic panels used to infill the cable net.

Cable nets have found more contemporary application in tensioned fabric roofs, as well, including the roof of the Munich Ice Rink located near the Olympic roof. The design (shown on the next page and discussed on pages 41 and 42) illustrates the fundamental reasons for using a cable net. The profile of the arch is fairly low relative to its span, so that fabric curvature is minimal. Because of the strength and stiffness provided by the net, however, the roof still behaves reliably. The net figures prominently in the aesthetic ideals conceived for the building by engineer Jörg Schlaich, as well. "It gives a scale to the structure," he notes, and serves to articulate a form that would otherwise be ill-defined from the interior. For Schlaich, use of the net represents less a practical choice than a philosophical one. "The main advantage of cable nets is the freedom in shape they give, and we should carry it as far as we can. The

The cable net provides visual definition to the subtly curving form of the Munich Ice Rink.

Source: Jörg Schlaich; used with permission.

membranes are not inexpensive, so we spend a little more to do them right" (Schlaich 1991, personal communication).

The effect of an interior cable net may be apparent on the exterior, as well. In single-arch or parallel-arch roofs, as well as in many tents, a tear in the fabric may result in a dangerous instability in the supporting structure. Because of this, designers often add external guy cables to provide stability independent of the fabric. Such cables often detract from the appearance of the structure by taking away from the drama of the curved fabric form. By tying the arch to the cable net via the vertical suspenders at Munich, however, Schlaich provided stability to the arch independent of the fabric. Other designers have taken advantage of the stabilizing effects of the cable net on conical roofs.

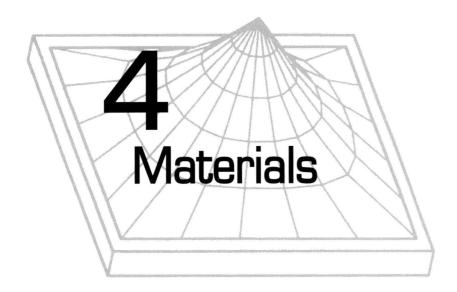

4 Materials

No built form has evolved more rapidly over the past 30 years than the tensioned fabric roof. A genre typified by crudely shaped and unreliable tents has been supplanted by one encompassing some of the world's longest free spans and most demanding architectural applications. These advances have been fueled partly by the development of nonlinear computer structural analysis techniques that permit accurate shaping and prediction of behavior under load for structures of rapidly evolving size and range of form. Developments in materials, however, have been the primary force behind dramatic improvements in durability, reliability, fire safety, daylighting, and energy use.

Three characteristics were common to traditional tents: portability, impermanence, and low cost. Throughout history, tent users have exploited their tents' light weight and flexibility by taking them with them as the animal herds and foreign armies they pursued or the entertainment they sponsored moved from place to place. As the tents deteriorated from hard use or weathering, their skins were replaced with new hides or fabric and used again. The materials used and the applications and sophistication of the tents improved, but their primary characteristics remained unchanged.

With the evolution in materials technology in the latter half of the twentieth century, however, and the introduction of fiberglass fabrics coated with polytetrafluoroethylene (PTFE) in 1973, the fundamental characteristics of fabric construction have been turned on their head. PTFE-coated fiberglass remains, overall, the highest quality fabric available, and the need for portability or low cost does not figure into decisions to use it. Where outstanding durability is required, however, along with dramatic form, daylighting, and good fire resistance, contemporary fabric structures offer unprecedented possibilities.

New materials will be required in order to further develop the market for tensioned fabric roofs, and it is surprising that none has attained a strong place in the market since Chemical Fabrics Corporation of the United States introduced PTFE-coated fiberglass 30 years ago. The material has undergone only modest evolution over the years, despite the fact that other manufacturers now market PTFE materials. Chemfab's John Effenberger, an expert in the material since the early years, believes that his firm's current product line, not radically different from that of 20 years ago, has not been surpassed by any of the new manufacturers (Effenberger 1998, personal communication). Taiyo Kogyo's Moto Nohmura, in discussing the material produced by a newer Japanese manufacturer, notes with some frustration that "the material is very similar to Chemfab's," and includes in his wish list for an improved material "a less expensive fabric of the same quality, and one with more control of light transmission" (Nohmura 1992).

Each of the material types currently available has a secure and almost mutually distinct range of uses. Important disadvantages accrue to each, however, even in their most favorable applications, and it remains to materials engineers to develop and successfully market improved products in the future.

The characteristics of fabric materials are fundamental to their applicability and success in the construction marketplace, and form the subject of this chapter.

Architectural fabrics are typically woven materials in which small orthogonal bundles of fibers are interlaced to provide a load-bearing "scrim" (see the drawings on the next page), upon which coatings or laminations are applied (Huntington 1987). The primary function of coatings is to provide an impervious, watertight finish. In addition, depending on the characteristics of the scrim material, coatings may reduce soiling of the material, protect it from abrasion, and provide protection from ultraviolet radiation or other damaging elements. There are a variety of materials used in tensioned membranes that do not employ the scrim-and-coating construction characteristic of most architectural fabrics. Some of these are discussed briefly in Chapter 3 and later in this chapter.

The behavior of a conventional fabric is in part a function of the particular weave employed. In plain weave, the simplest and most common, fiber yarns pass alternately above and below fibers in the perpendicular direction. While fibers in the warp direction of the weave are held taut during manufacture and remain nearly straight, those in the fill direction crimp around the warp fibers. The construction results in rather different behavior in the two directions, as the straight warp fibers give the fabric greater stiffness in that direction.

In the fill direction, the fibers must be stretched out straight, transferring their curvature into the warp fibers, before the fabric

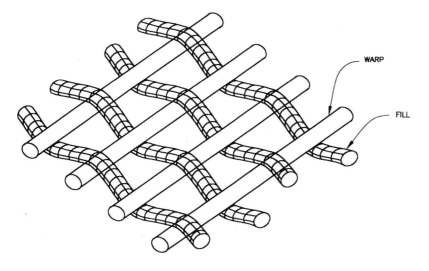

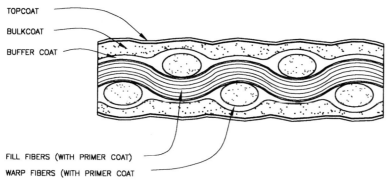

[ABOVE LEFT] In a typical plain woven fabric, fill fibers are crimped about the straight warp fibers. The fill fibers straighten out as they pick up load, making the fabric relatively stretchy in this direction. The material is fabricated in rolls with the warp direction running lengthwise.

[LEFT] Coatings protect the fibers from abrading each other under load.

begins to stiffen and carry significant load. This phenomenon is termed "crimp interchange," and it has the adverse effect of causing fill fabrics to "creep" into a straighter profile under prestress, thereby losing some of their initial tension. In addition, crimp interchange places accurate stress analysis beyond the capability of currently available computer software, which generally assumes an unvarying relationship between the stress in a material and its strain (deformation). Some manufacturers have sought to address these problems by using laid weaves (fibers are placed one atop the other in the two directions without interweaving); stitch weave (a laid weave with the two directions "stitched" together by a separate thread at selected intervals); or panama weave (where the material is interwoven at every second, third, or fourth fiber). As an alternative, the weaver may hold both warp and fill fibers in tension during weaving so that the material has similar stiffness properties and limited crimp interchange in both directions.

Fabric Performance Parameters

Performance parameters of architectural fabrics include their mechanical properties, durability, light transmission, fire resistance, and cost, as described below.

Mechanical Properties

The critical mechanical properties of architectural fabrics are those related to tensile strength, tear strength, and stiffness. Tensile strength tests measure the level of direct pull force required to rupture the fibers of the material, and hence give a measure of the fabric's ability to resist the tensile forces resulting from prestress in combination with external loads.

Tear strength provides a measure of the resistance that a fabric that has been slit or cut has to propagation of the tear, and hence of its ability to prevent localized overstress or damage from resulting in broader tears. The correlation between tensile and tear strength is low, as stiffly coated fabrics such as PTFE-coated fiberglass may not have the flexibility to allow undamaged fibers to bunch or "rope" at the end of a tear in a manner that prevents the tear from advancing.

The stress/strain or stiffness behavior of a fabric is a complex phenomenon that is not reducible to a single variable. Stiffness is of course related to the modulus of elasticity (stress divided by strain) of the material and the area of fibers employed, which may vary in the warp and fill directions of the material. In addition, the type of weave employed and the manufacturing process will both affect stiffness variation under load due to crimp interchange. The effect of crimp interchange on the stress/strain relationship is a function of the tension applied along both axes of the fabric. By applying high tension to the warp fibers, for example, the tendency of the fill fibers to straighten out under load is reduced, and the fabric's stiffness in the fill direction is increased (see the drawing on the top of page 59). As noted earlier, analytical techniques currently in common use do not provide explicit consideration of the tendency of fabric stiffness to vary with stress in both the direction under consideration and the opposing direction. Nonetheless, understanding of the complex stiffness properties of architectural fabrics is useful to the design and construction of reliable fabric roof structures.

A number of tests have been developed in the United States, Europe, and Japan to model the tensile strength, tear strength, and stress/strain properties of fabrics (see the drawings on the next page). Such test standards are periodically updated, and designers are advised to confirm current requirements in establishing specifications for a particular project. At the time of this writing, major changes are underway in the Japanese Industry Standards (JIS), as published by the Japanese Standards Association (JSA), and new European Neutral (EN) standards are under development. The most common and useful of current standards are described below.

Strip Tensile Test. The tensile strength of uniaxially loaded fabric is typically measured in the United States by the American Society for Testing and Materials (ASTM) Standard D4851 (ASTM 1997) in

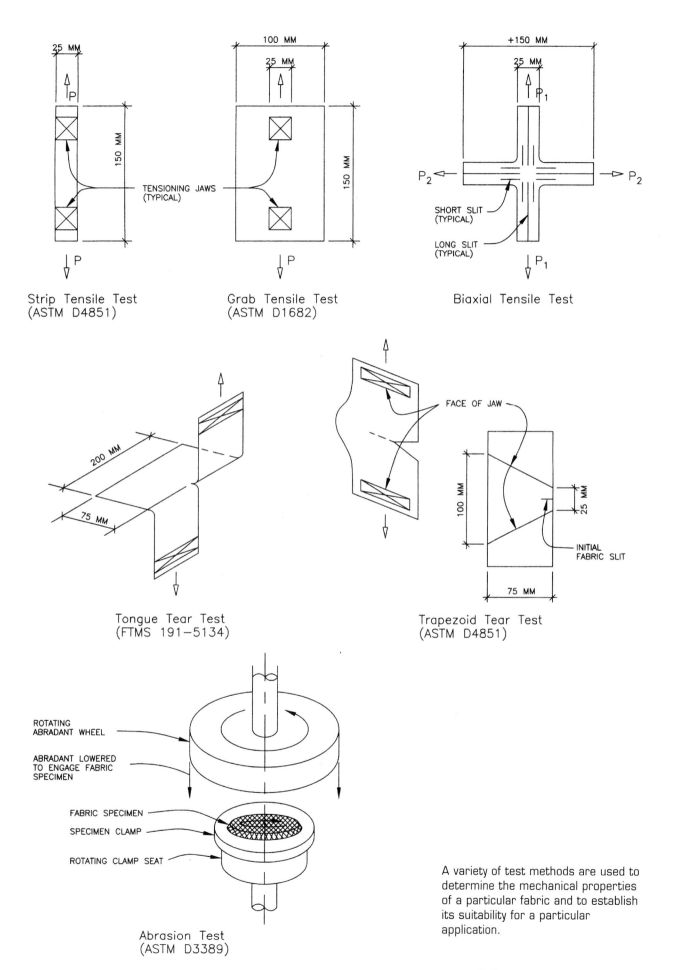

A variety of test methods are used to determine the mechanical properties of a particular fabric and to establish its suitability for a particular application.

which a strip of fabric 25 mm in width is held by clamps at both ends spaced 75 mm apart and pulled at a rate of 50 mm/min until the fibers rupture. Equivalent tests are provided by European standard DIN 53354 (DIN 2002) and JIS L1096 (JSA 1999a). Refinements of the standard strip tensile test that are employed by certain manufacturers or defined in project specifications include tests of the material in a pre-wetted condition or after folding it. ASTM D4851 provides standards for both test variations, and the Membrane Structures Association of Japan (Ishii 1995) has dictated that fabric shall retain 80 percent of its strength after 72 hours of water immersion, 2000 hours of weatherometer testing, or 5 years of daylight exposure. Tests for the effects of folding, sometimes called "flex fold" (Birdair 1996), document the effect that fiber damage due to creasing has on tensile strength. While strip tensile tests are the most commonly employed tests of fabric mechanical properties and are typically used as the basis for determining the allowable stress in fabric, several factors combine that may cause the test to significantly overestimate the usable strength of the fabric (Murrell 1984), as follows:

1. Fibers in the direction transverse to the direction of pull are very short (25 mm), so that tested fibers are able to straighten out to carry load most effectively during testing.
2. Defects in fibers or distortions in weave that tend to appear as weak links limiting the effective strength of fabric in a full-size roof are unlikely to be present in a fabric sample of the size used for testing.
3. Punched holes at fabric clamps and angle changes at bearing points may cause localized stress concentrations, leading to premature failure.
4. Test samples, unconstrained by the complex geometries and fabrication considerations of actual structures, are carefully cut so fibers are aligned in the test direction.
5. Fabrics typically fail by tearing rather than rupturing in pure tension, and the tear strength of fabrics with similar strip tensile strength can vary dramatically.
6. Ultraviolet radiation and other factors can cause degradation of tensile strength over time, with significant losses occurring over the course of the useable life span of some materials.

Grab Tensile Test. A test less commonly employed than strip tensile is the grab tensile test, performed in the United States in accordance with ASTM D1682-59T (ASTM 1975) or Federal Test Method Standard 191, Method 5100 (FTMS 1989). The strip and grab tensile tests deviate in the width of material tested. While the full 25-mm width of fabric is gripped by the jaw in the strip tensile test, in the grab tensile, the 25-mm-wide jaws grip the center of a 100-mm-wide strip of fabric, so the material's ability to redistribute high local stresses is tested.

Biaxial Tensile Test. In order to document stress/strain behavior under varying combinations of load in the warp and fill directions, the test sample is cut into a cruciform shape, so that pairs of jaws can pull simultaneously in the two directions. Biaxial tensile tests are performed by manufacturers in order to provide complete documentation of the stress/strain behavior of their materials for use in patterning, but are not done under the purview of any standard test method.

Seam Tensile Strength Tests. The usable tensile strength of a membrane in a direction perpendicular to its seams is limited by the strength of those seams, and both strip and grab tensile tests have been adapted for use in testing this property. ASTM D1682-59T grab tests are designated D1683-90a (ASTM 1990) when performed on sewn seams. It is important that tests of seams evaluate both creep under sustained load and performance at elevated temperature, as service temperatures on dark-colored fabrics in direct sunlight can rise to as high as 65°C. There are no universally accepted standards for seam strength. The Membrane Structures Association of Japan, however, dictates that welded seams tested in accordance with JIS L1096 shall have a strength equal to 80 percent of the base fabric and sewn seams 70 percent. The American Society of Civil Engineers (ASCE) has developed a standard for air-supported structures requiring seam strength equal to 200 percent of calculated fabric stress at 20°C and 100 percent of calculated fabric stress at 70°C, with test loads sustained for 4 hours (ASCE 1996). ASCE's soon-to-be-adopted standards for fabric tension structures will add to the above requirements the stipulation that seams shall resist a load equal to 100 percent of the fabric's minimum specified tensile strength.

Trapezoidal Tear Test. The commonly employed method for determining the resistance of fabrics to the propagation of tears is the trapezoidal tear test. U.S. specifications typically reference ASTM D4851. The Japanese standard JIS L1096, British Standard 3424 (Method 7B) (BSI 2000), and the Deutsches Insitut für Normung e.V. (DIN) 53363 (DIN 2000) provide equivalent tests. In performing the test, a short cut is made along one edge of the fabric, perpendicular to the direction of pull, and the fabric is placed between the jaws of the testing device with the cut edge held taut and the other slack.

Tongue Tear Test. By testing the fabric in its own plane, the trapezoidal tear test provides the tear test method most directly applicable to architectural fabrics. In the tongue tear test, performed in accordance with Federal Test Standard No. 191-5134 (FTMS 1978) or British Standard 3424 (Method 7C) (BSI 2000), the sample is split over a portion of its length, with one of the resulting "tongues" pulled upward and the other downward by the testing device jaws.

Adhesion Test. Adhesion tests provide a measure of a fabric's resistance to the peeling off of its coating, and may be performed in accordance with ASTM D4851 or British Standard 3424 (Method 913) (BSI 2000). They are most useful in measuring the resistance to peeling apart of cemented or heat-sealed seams.

Abrasion Resistance. In the United States, resistance to abrasion may be tested in accordance with ASTM Standard D3389 (ASTM 1999c), while Japanese testing is performed in accordance with JIS K6328 (JSA 1999b).

Durability

The durability of a fabric membrane is a complex performance parameter dependent on resistance to ultraviolet radiation degradation, damage resulting from wicking (the absorption of water along the length of yarn fibers due to capillary action), attack from algae or other organic matter, and the retention of seam strength. In addition to the limits imposed by these sources of structural damage, the practical life span of architectural fabrics also may be limited by permanent stains from soiling, reductions in the translucency of the material, or vandalism. Vulnerability to these hazards varies widely among materials. The effect on each is discussed in later sections of this chapter.

The most reliable measure of a material's durability is established through evaluating the performance of built structures. Both polyester coated with polyvinyl chloride (PVC) and PTFE-coated fiberglass, the two most commonly used architectural fabrics, have an ample record of performance to establish their durability under varying environmental conditions.

For newer materials or products employing altered material or manufacturing specifications, accelerated weathering tests may be performed to establish long-term resistance to degradation from sunlight and water. The most common of these is done using a carbon-arc light source in combination with an intermittent water spray. This test is performed in the United States in accordance with ASTM Recommended Practice G153-00ae1 (formerly G23) (ASTM 2000b). There are no adopted acceptance criteria for weatherometer performance or widely accepted correlation between weatherometer and exterior exposure periods, although one industry source (DuPont

1983) has correlated 5,000 weatherometer hours with 15 years of exterior exposure. The Membrane Structures Association of Japan has advised that materials other than PTFE-coated fiberglass should retain 80 percent of their tensile strength after either 2,000 weatherometer hours or 5 years exterior exposure.

Resistance to weakening of the fibers from long-term ultraviolet radiation exposure is the most commonly noted factor in fabric durability. The life of some materials also may be shortened by wicking, which can lead to freeze-thaw damage of the material in cold weather conditions. Wicking also may provide a moist atmosphere for mildew growth, a source of deterioration in both fiber and seam strength and a possible cause of permanent discoloration. Wicking is best prevented by the provision of adequate coating thickness over the entire area of the scrim, as well as the design of structures that are adequately sloped throughout the membrane, that have seams laid out to "shingle" water away from the cut ends of the material, and that are designed to avoid condensation. Mold and algae growth may occur on fabric surfaces as well as in the fabric interstices through wicking, and should be prevented by the same measures as described above.

Damage from vandalism also can effectively limit the durability of tensioned fabric structures, and attention must be given to its prevention, although relatively few structures built to date have had problems. In an incident at Spencer Secondary School in Langford, British Columbia, youths caused $40,000 damage with a knife in June 1982 (Sidenius 1982). The University of La Verne Student Center in Southern California also has been subject to more abuse than most. College pranksters shot an arrow into the fabric, and vandals managed to cut several knife holes in the fabric at locations where it descends to within 1 m of the ground. Holes from the knives and the arrow were all small, however, and none propagated into a rip. They were readily repaired by college maintenance personnel using fabric patches and a special iron.

The La Verne experience demonstrates the advisability of shaping the tents so that they cannot be reached from ground level, particularly when they are located in unsafe areas or on unsecured sites. Significantly, La Verne's director of maintenance notes that there has been less damage from vandalism at the Student Center than at other college buildings. Whether this is due to the affection the community has for its unusual structure or something else is not clear, but it should alleviate concern over the potential for vandalism on other fabric structures. In terms of material selection, provision of adequate tear strength may be the most important factor in limiting vandalism damage, so that small areas of damage do not propagate into more global failures.

The effective life span of a fabric membrane also may be limited by permanent and unsightly soiling, staining, or loss of translucency.

While such damage typically has an accidental cause, vandalism has been a source of irreparable damage as well. Resistance to dirt and staining is primarily a function of the properties of the coating material, including the topcoats that are used with some materials. Unfortunately, the "softness" of coatings such as PVC, which gives fabrics good tear strength and resistance to damage during handling, also contributes to their tendency to be less resistant to soiling.

While field use provides the best measure of a material's resistance to soiling, certain standardized tests have gained limited use as measures of stain resistance and cleanability. In the United States, the ASTM D1308 (ASTM 2002) test may be employed for stain resistance, using staining materials such as acids, bases, organic solvents, motor oil, and mustard. ASTM C-756 (ASTM 1999b) is used to measure cleanability with soiling agents such as asphalt, tar, and paint (DuPont 1983).

Light Transmission

A fundamental aspect of the design and appeal of most tensioned fabric structures is their translucency, or ability to transmit a portion of the light that strikes their exterior surface into the interior of the space. The fiber bundles in a structural fabric have a density that makes them nearly opaque, and translucence is therefore primarily a function of the translucency of the coating material and the size and spacing of the gaps between the fibers. Translucence is generally measured as a percentage of solar energy striking the exterior fabric surface that is transmitted into the space and may vary from near zero for certain heavy, dark-colored fabrics, to 95 percent or more for greenhousing films.

Of the light that is not transmitted, some fraction is reflected back outward, while the remainder is absorbed into the material and reradiated as heat either back out or into the space. These values also may vary widely, with white fabrics generally high in reflectance and dark fabrics high in absorption. In the United States, light transmission properties may be measured in accordance with ASTM E424 (ASTM 2001b).

In real structures, response to light is more complex than the three percentages for transmission, absorption, and reflectance defined by test values, as these percentages may vary for different light wavelengths, and multiple fabric layers or insulation may dramatically alter light behavior. In addition, response to light constitutes only one factor in the energy behavior of a fabric-roofed building. These considerations are discussed in more detail in Chapter 8.

Fire Resistance

Contemporary tensioned fabric structures have the ability to provide fire safety far better than that of traditional nonsynthetic tenting materials. Testing over the last 30 years has demonstrated the excellent fire resistance of glass fiber fabrics, and more recently developed silicone-coated fiberglass fabrics have a fire resistance approaching that of PTFE-coated fiberglass (Fabric Structures International 1997). Several standard fire tests have been adapted for use in measuring the fire performance characteristics of these materials. The model codes have begun to recognize and require fire testing of membranes in accordance with the following tests:

1. **ASTM E84, Surface Burning Characteristics of Building Materials (known as the Flame Spread Test) (ASTM 2001a).** The test is applicable to exposed surfaces such as ceilings or walls, and measures surface flame spread and smoke development relative to mineral fiber cement board (index of 0) and select grade red oak flooring (index of 100). Building codes limit smoke generation to an index of 450 and categorize flame spread as Class I (0-25), Class II (25-75), and Class III (75-200).

2. **ASTM E108, Fire Tests of Roof Coverings (Tests known as burning brand, spread of flame, intermittent flame exposure, flying brand, or rain test) (ASTM 2000a).** These tests evaluate roof coverings to measure their resistance to fire originating outside the building. Class A tests are applicable to roof coverings that are effective against severe test exposure, Class B tests are applicable to coverings that are effective against moderate exposure, and Class C tests are applicable to coverings that are effective against only light exposure. In all cases, the covering must not slip from position or present a flying brand hazard.

3. **ASTM E136, Behavior of Materials in a Vertical Tube Furnace at 750°C (ASTM 1999a).** This is a test of base fabric material (also called "greige goods") and is not intended to apply to laminated or coated materials. A $1\frac{1}{2}$-in.^{2} stack of material is placed in the furnace to verify that the temperature of the material does not rise more than 30°C above that of the furnace and that no flaming from the specimen occurs after the first 30 sec.

4. **NFPA 701, Fire Tests for Flame-Resistant Textiles and Films (NFPA 1999).** The test method determines the difficulty of igniting flame resistant textiles and films and the difficulty of propagating flame beyond the area exposed to ignition. Small- and large-scale tests evaluate resistance to small and large ignition sources.

Cost Considerations

The remarkable properties of fiberglass and other synthetic fabrics have enormously expanded the range of fabric tension structure application, as discussed throughout this chapter. Portability, one of the two most salient features of the old tents, has been lost in the bargain. So, too, has the other: extreme economy in enclosing space.

Small size, complex shapes, lack of symmetry, and the use of liners are all factors that tend to raise the cost of a structure relative to its plan area. Complex interfaces with surrounding elements, such as irregular attachment of the membrane to an adjoining wall or roof, can increase the complexity of detailing and the difficulty of installation, and thereby cause significant increases in cost, as well.

Structural engineering costs for fabric roofs are high relative to most forms of construction due to their complex analysis and detailing requirements. For most structures, engineers will accrue fees of 6 to 12 percent of construction cost for the design of cabling, supporting elements, and connections. The development of fabric patterns and fabric detail drawings (analogous to but much more complex than the drafting of steel or rebar shop details) may cost an additional 4 to 6 percent of construction cost.

Certain applications, however, remain economical in fabric. Air-supported fabric roofs remain the most economical means of attaining clear spans in excess of 100 m, while fabric-sheathed cable domes are competitive in cost with conventional rigid roofs. Similarly, when fabric is used as a substitute for a rigid glass or fiberglass skylight where daylighting is required, as in a shopping mall, it is often cheaper than competing systems.

Simple construction cost comparisons cannot, however, take account of some of the intangible value of fabric roofs to their owners. Leonard Harper, former coordinator of student activities at La Verne, called the building "our showpiece," and indicated that "the University would have real attrition if it weren't for the Campus Center" (Harper 1986, personal communication). Harper's statement, which echoes the feelings of many who own or work under a fabric roof, infers both the raw visual excitement of the structures and their effectiveness as symbols or signs of an institution.

PVC-Coated Polyester

Fabrics that employ coatings or laminations over a polyester substrate are both the oldest and the most commonly used materials in the contemporary fabric structure palette. Polyester substrates are typically tightly woven and imperviously coated, although open weaves or light, open knits have sometimes been employed on decorative or shade canopies (shown on the next page). In some lower grade fabrics, impervious PVC layers are adhesive laminated to the substrate, while most are PVC coated. Other coating materials such as neoprene and hypalon have been employed on radar domes and for industrial fabrics, where opaqueness and recoating in the field are required. Urethane coatings are useful for materials that must be

Coated polyesters are available both in conventional weaves with waterproof coatings (left) and as open-woven shade fabrics (right).

folded and unfolded in temperatures cold enough to cause cold cracking in other materials. Those alternative coatings are rarely employed for architectural applications, however.

Polyester fabrics with PVC coatings or laminates came into wide use in the 1960s. These synthetic materials proved considerably stronger than cotton or other natural fibers; they are not subject to rot and their improved ultraviolet radiation resistance gives them greater durability, as well. PVC-coated polyester fabrics have become the mainstay of the industry and have proven their worth in structures that often are well-shaped and well-constructed, with reliable long-span capability. Even today, more tensioned fabric roofs are built with this material than any other, and they dominate such applications as high-profile air-supported tennis-bubble–type structures, portable rental tents for shows or events, and park shade canopies or amphitheater covers.

PVC-coated polyesters retain the two most desirable features of traditional tent construction: they are relatively inexpensive and they have a flexibility and ease of handling that make them suitable for the repeated erection and disassembly required of portable structures. They have not, however, entirely eliminated certain of the drawbacks of traditional materials, as their life span of 10 to 15 years may be judged marginal for a "permanent" material, and they have difficulty in passing certain fire resistance tests. Building code and safety issues sometimes limit their use for permanent, enclosed spaces, and they may not be the first choice for certain high-profile applications.

The materials are available in a wide range of colors, although darker colors have dramatically reduced translucency and are more subject to fading.

The performance parameters of PVC-coated polyester are described below.

Mechanical Properties

The polyester base material has fiber tensile strength ranging from 350 to 1,200 MPa (Textile World 1970). Using structural membranes with a weight of 800 to 1,100 g/m^2, strip tensile strengths of 3,100 to 5,800 N/5 cm can be achieved (Seaman 1984), which is adequate for most applications. PVC-coated polyester fabrics also are noted for having good tear strength, as the relatively soft coating material allows unbroken fibers to rope at the end of a tear in a manner that strengthens it against further damage.

Polyester fabrics generally have moderate stiffness and a moderate tendency to creep under load that necessitates retensioning of some structures. These properties also create advantages for the material, however. Because its stiffness is low relative to that of fiberglass products, compensation (a measure of the amount that the membrane must be undersized in fabrication in order to be appropriately "pretensioned" in its final shape) is correspondingly high, so that small errors in patterning or fabrication are less likely to result in overstress or wrinkles. Creep, furthermore, results in a beneficial relaxation of areas with excessive prestress.

Some of the dimensional instability of the fabrics is a result of crimp interchange between the relatively straight warp fibers and the fill fibers that weave around them. Seaman Corporation of the United States has addressed this by employing a laid weave, so that no crimp interchange occurs when the fill is tensioned. This approach has the further advantage that, by providing a flatter texture to the scrim, the weight of PVC required to effectively coat the fibers is minimized. Ferrari of France manufactures fabric with a more conventional plain weave, but holds both warp and fill in tension during the weaving process so that the fabric has equal stiffness in both directions and crimp interchange is again minimized (Ferrari 1999).

Durability

While polyester may be degraded by exposure to ultraviolet radiation, the PVC coatings provide protection adequate to provide a 10- to 15-year life span in most exposures. These materials also may be degraded by prolonged exposure to animal fats, as may be contained in foods or in soap (shown on the next page). Some polyester fabrics are manufactured with an additional "top" coating on their upper surface. Seaman employs a polyvinyl fluoride (PVF) topcoat with the trade name Tedlar, while Ferrari and others use polyvinylidene fluoride (PVDF) with trade names such as Kynar® or Fluorex®. While PVDF impregnates the PVC through the application of heat and pressure, PVF is applied as a film using adhesives. Advantages accrue to each. PVF laminates avoid the small pinholes that occur in

Close proximity to a fast-food restaurant left this PVC-coated polyester canopy vulnerable to damage from animal fat exhausts. Other canopies only a short distance away were undamaged.

coatings and that may lead to bleedthrough of the PVC undercoating and eventual soiling. PVDF coatings, on the other hand, are not subject to the delamination that has occurred with certain laminated topcoats (Dery 1998). Acrylic and urethane topcoats also are used.

While topcoats provide additional ultraviolet (UV) protection, thereby yielding a moderate improvement in fabric life span, their primary value is in the resistance to dirt adhesion and improvement in cleanability that they offer. These result from increases in resistance to staining and plasticizer migration, the process by which the various compounds applied to the fabric are leached out by exposure to sunlight. The effect of plasticizer migration is similar to that of the gummy residue left when a sticker is removed from a car dashboard (Grossman 1991).

Surfaces topcoated with PVF and PVDF cannot be seamed using conventional radio frequency welding. Because of this, fabricators must use abrasives or solvents to remove the topcoat from edges to be seamed. Fabric manufacturers can provide material with the topcoat left off one edge of the fabric roll, however, so that topcoat removal is not required where unpatterned (straight) fabric edges can be used.

Along with the direct impact of topcoats on material cost and the labor associated with topcoat removal at seams, contractors have on occasion been dissuaded from using topcoats because the increase in fabric stiffness that they cause may lead to a modest increase in the difficulty of handling the material during fabrication, patterning, and erection.

Light Transmission

White structural weight polyester fabrics with PVF or PVDF topcoats are able to attain translucency of up to 22 percent. Accumulated dirt or long-term discoloration may reduce translucency levels, however, and colored materials have lower translucence. In some applications, such as arena shows requiring purely artificial lighting, opaque materials are preferred. This can be achieved with polyester fabrics that employ black film. The film can be sandwiched between white coatings to provide a material that is both white and opaque (Bradenburg 1999, personal communication). While such fabrics sacrifice the energy savings associated with interior daylighting, the white exterior coat maintains the high reflectivity required to moderate heat gain in warm weather.

Fire Resistance

Polyester fabrics will begin to melt at temperatures in excess of 160° to 250°C (Lambert 1986); but the fabrics perform well in ASTM E84 spread-of-flame tests, with the ability to achieve passing smoke generation indices and Class I flame spread ratings. The fabrics extinguish upon removal of the heat source. In practice, melting of the fabric under high temperatures may create holes in the membrane that allows dissipation of heat and smoke, contributing to the safety of their use. The materials are generally also able to pass small- and large-scale tests of NFPA 701. At present, however, the performance of polyester fabrics on ASTM E136 tests generally prevents their classification as noncombustible in the United States, although their NFPA 701 performance has earned them the designation "flame retardant."

Cost

The cost of unfabricated polyester is the lowest of the commonly used membrane materials, and the relative ease of seaming the material and handling it during fabrication and erection further contribute to its low installed cost. While fabric weight and topcoat affect cost, the 2003 price in the United States for unfabricated material is approximately $12/m^2. The cost of a fabricated membrane may range from $90 to $150/m^2, while a complete polyester roof structure (including fabric, cables, clamping, and supporting structural elements, but excluding foundations or a building on which the roof rests) typically will cost in the range of $400 to $700/m^2 of plan area. In considering these costs, it should be remembered that approximately 20 to 25 percent of the area of unfabricated material is lost to seams, cuffs, or waste in the fabricated membrane and that, due to slopes in the roof, the plan area of the completed structure is typically 75 to 90 percent of the area of the fabricated membrane. The combination of moderate cost and generally good performance has made PVC-coated polyester fabrics the workhorse of the industry.

PTFE-coated fiberglass is manufactured in both liner (left) and structural weights (right). The beige pigment of the structural fabric will fade to white after extended sunlight exposure.

PTFE-Coated Fiberglass

Developments beginning around 1969 were to fundamentally alter the character of fabric roof construction. At that time, the DuPont company teamed with Owens-Corning, Chemfab, and Birdair, Inc., to create a fiberglass fabric coated with Teflon® resin that was to become the most important advancement in membrane construction since the development of canvas (shown above).

The new fabric was first used at the University of La Verne Student Center in 1973, where its incombustibility, resistance to soiling, and projected durability brought fabric structures entrée to a broad range of new, high-end architectural applications. At the same time, the high cost of the material and its susceptibility to damage from the rough handling of repeated erections shut it out from the traditional uses of fabric in low-cost, portable tent applications.

The following paragraphs describe the performance parameters of PTFE-coated fiberglass.

Mechanical Properties

The yarns that form the base of the fabric have a tensile strength of about 3,500 MPa, greater than any commercially available steel (Textile World 1970). The strength of polyester yarns, by contrast, ranges from 350 to 1,200 MPa. The strip tensile strength of finished PTFE-coated fiberglass materials may range from about 1,600 N/5 cm for light liner materials to 8,800 N/5 cm for the heaviest structural membranes (Chemical Fabrics Corp. 1995). The high strength of glass fiber permits its use in structures with long spans and minimal curvatures.

Fiberglass is a stiff material, and materials with scrims of fiberglass deform less under load than those using polyester. In the

THE TENSIONED FABRIC ROOF 73

process of applying coatings to fiberglass, warp fibers are held in tension. As the freshly applied PTFE (applied at temperatures in excess of 320°C) cools, it shrinks, drawing crimp into the fill fibers while those in the warp remain straight. The resulting material is characterized by high warp stiffness and fill stiffness that is relatively low under low stress but that increases as crimp interchange occurs under higher stress (Geiger 1989).

The Achilles heel of PTFE-coated materials has been their limited tear strength, which is restricted by the inability of the relatively stiff coating to permit fibers to bundle at the end of a tear in a manner that stops its propagation. Typical trapezoidal tear values range from 80 N for liners to 550 N for heavy structural membranes (Chemical Fabrics Corp. 1995).

The sealing of PTFE-coated fabrics is generally achieved by the heat welding of lapped fabric edges with a Teflon® adhesive intermediate strip, usually about 50 mm in width, and has been demonstrated to reliably develop the full strength of the base material. Because the FEP has a lower melting temperature (200°C) than the PTFE coatings (260°C), it can be effectively sealed to the base fabric without damage to the coatings (Geiger 1989).

Durability

Since its first use at La Verne, the developers of PTFE-coated fiberglass have projected a minimum 20-year life span for their product, and the accuracy of this prediction of "permanent fabric architecture" has been the critical factor in the success of this material in the marketplace. In point of fact, La Verne continues to perform reliably 30 years after its construction, and incidences of PTFE-coated fiberglass being replaced due to wear or degradation are almost nonexistent.

The material does, however, have several vulnerabilities. Raw fiberglass fibers are subject to damage from the wicking of moisture into the exposed edges of the material, which may damage the fabric substrate. By treating the yarns with water repellents, however, manufacturers have generally avoided degradation (Dery 1998, personal communication). In addition, substantial variations in coating thickness over the irregular surface of the fiberglass scrim result in imperfections in the fabric's "hydrophobia" that can lead to damage from limited water infiltration through the body of the material (Dery 1992). The material's low tear strength creates a potential limitation on its durability, as well, by increasing its vulnerability to the propagation of rips initiated by vandalism, accident, or overloading.

Vinyl-coated polyesters, as noted earlier, are relatively easy to handle and resistant to damage from creasing or folding. Fiberglass fibers, however, are inherently brittle and subject to fracture if handled roughly. Although serious damage is rare, the fabric still must be handled carefully by experienced workers to ensure proper erec-

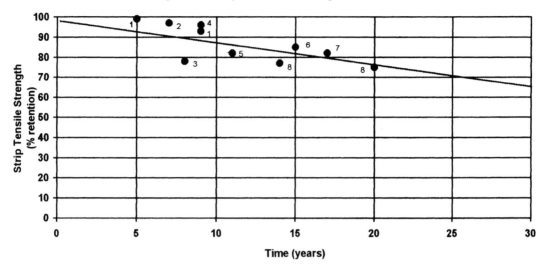

Projected Strip Tensile Strength Retention

tion. The care required in its handling prevents the use of PTFE-coated fiberglass in structures that are moved or dismantled seasonally, but the strength, durability, and relatively high cost of the fabric make it best suited for permanently sited structures anyway.

Test data on fabric samples taken on structures of up to 20 years of age have borne out the predicted durability of the material (see chart above). Accumulated test data from structures of varying ages and exposures appears to indicate that strength degrades in approximately linear fashion, with La Verne, tested at 20 years of age, retaining 75 percent of its original strip tensile strength (Birdair 1996).

The critical factor in the durability of the material is its near invulnerability to degradation from UV radiation. The fabric's PTFE coatings, applied in several layers, are the key to this remarkable durability. PTFE and the glass fibers themselves are both highly resistant to chemical attack, UV radiation, and wide temperature variations. The material has unsurpassed durability under conditions of extreme sunlight, heat, cold, and air pollution. Although algae may grow on dirt accumulations in structures that go through regular wet/dry cycles, the PTFE coating itself will not sustain algae growth, which can be removed simply by cleaning the dirt from the fabric (Dery 1998, personal communication).

During their erection, structures like La Verne are usually stretched to initial prestress levels of 150 to 350 N/5 cm throughout the roof. In 1979, tests were made by Owens-Corning Fiberglass to determine the levels of stress in the La Verne fabric after more than 6 years of use. They indicated stresses falling in the range of 230 to 325 N/5 cm with an average of about 265 N/5 cm (Owens-Corning 1979). The relatively uniform stress level, at or near the original pre-

Tests have demonstrated the very slow degradation of strength in PTFE-coated fiberglass fabrics in various applications and climates.

Note: 1, RCA Dome, Indianapolis, IN; 2, Atrium Court, Newport Beach, CA; 3, Florida Festival, Orlando, FL; 4, Franklin Park Zoo, Boston, MA; 5, Dakota Dome, Vermillion, SD; 6, UNI Dome, Cedar Falls, IA; 7, Santa Clara University, Santa Clara, CA; and 8, La Verne College, La Verne, CA

stress level, demonstrated that the fabric had not stretched or distorted over time. This stretching, generally referred to as "creep," is only about one-third as high in glass fiber as in polyester fabrics. Because most glass fiber creep occurs in the first 2 to 3 days under stress, its effects can generally be compensated for over the period of time required for full prestressing of the membrane. The uniformity of stress levels at La Verne attests to the fabric's low creep, in addition to verifying the accuracy of the original computer analysis performed by the structural engineers.

Light Transmission

Commercially available structural weight fabrics of plain weave construction generally have translucence in the range of 7 to 21 percent, with liner membranes ranging up to 27 percent. By using more complex weaves, however, the interstices between fiber bundles can be increased in size, and structural fabrics with translucence of 23 to 25 percent are now on the market (Dery 1998, personal communication).

PTFE-coated fiberglass has typically been supplied only in white, and, while colored fabrics are now available, their translucence is limited to 1 to 3 percent.

Fire Resistance

In general, contemporary fiberglass-based fabrics are able to satisfy the noncombustible rating requirements of the building code.

The ASTM E108 test stipulates that the roof resist the heat from a burning wooden brand for several minutes without permitting any embers to burn through the fabric. Newer PTFE-coated fiberglass products are able to pass both the small-scale (Class B) and large-scale (Class A) burning brand tests of ASTM E108.

Cost

PTFE-coated fiberglass fabrics are high-technology products subject to strict quality control, and unfabricated rolled goods may cost $60 to $80/m^2 of fabric area. The additional costs associated with fabrication, the added fabric area created by the slope of the roof, the addition of cables, mast or arch supports, and field erection bring finished roof cost to $500 to $1,000/m^2, exclusive of foundations.

Silicone-Coated Fiberglass

A newer material developed by ODC, a firm created jointly by Oak Industries and Dow Corning, substituted coatings of silicone for PTFE in an effort to address the drawbacks of the older material: low tear strength, difficulty in handling, limited translucency, and high cost.

Silicone-coated fiberglass realized some improvements in these areas, but a number of structures developed problems with seam strength and dirt accumulation that raised concerns about the material. At present, the material has not found a secure position in the marketplace. Performance parameters are described below.

Mechanical Properties

Material strip tensile strengths are largely a function of the woven scrim and therefore similar for silicone and PTFE-coated products. The silicone provides a somewhat "softer" coating for the brittle fiberglass fibers, which proponents of the material believe prevents the fiberglass filaments from abrading each other when the material is creased or mishandled. Several portable or seasonally erected structures have in fact been built using silicone-coated fabrics. Others believe that the softness of the coating allows the fabric to bend to such a small radius when creased that fiberglass filaments are more easily broken. Flex fold testing appears to indicate that PTFE- and silicone-coated materials retain a similar percentage of their initial strength when creased.

While the softness of the silicone coatings allows undamaged fiberglass fibers to rope slightly at the end of a cut or tear in the fabric, testing to date has not demonstrated a significant improvement in tear strength.

Silicone coatings, unlike PTFE coatings, are not applied at temperatures that crimp the fiberglass fibers upon cooling. Because of this, the full stiffness of the fiberglass yarns is realized in the finished fabric, which has extremely high stiffness in both warp and fill directions that makes it unforgiving of errors in patterning or fabrication (Geiger 1989).

The builders of silicone-coated fiberglass structures had difficulty in creating seams between the strips of fabric within a structure, resulting in major seam failures on several structures. Unlike PTFE-coated fiberglass structures that use heat welding of seams, silicone-coated fiberglass seams must be made using adhesives, and some engineers prefer to limit the use of silicone-coated materials to designs where fill fiber (and seam) stress can be maintained at a low level (Huntington 1990).

Durability

The "softer" properties of the coating that improved its handleability and tear resistance also made it more susceptible to the accumulation of stains from dirt, pollution, or other sources. Some stains proved difficult to remove, and the initial pristine whiteness of the fabrics was difficult to maintain. Structural engineer Horst Berger mirrored the frustration of others with this property of the material in saying "the cleaning problem (with silicone) can be solved. All you need to do is find a way to close those little pores with something other than dirt" (Berger 1992, personal communication).

Light Transmission

Early materials had translucence of 35 percent or more with high-strength fabrics, while PTFE products of the time were limited to less than 20 percent. The materials provided daylighting approaching that of glass skylights and permitted healthy growth of nearly all plant species. The translucence of more recent silicone fabrics is limited to about 25 percent, however, due to the addition of fire retardant materials, while some PTFE fabrics are now attaining similar values.

Fire Resistance

Early silicone-coated materials were limited in fire resistance, but more recently developed products have performed well enough on ASTM E108 and NFPA 701 to attain the noncombustible rating previously limited to PTFE-coated fiberglass.

Cost

Both material cost and installed cost for structures built using silicone-coated fiberglass fall into a range generally below that of PTFE-coated fiberglass but well above that for PVC-coated polyester.

Films

Films are thin sheets of synthetic material used without benefit of a woven "scrim" substrate. Lacking a scrim, they are not actually fabrics. As they are used in applications similar to fabrics and share the membrane structural behavior of fabric, however, they deserve mention here. PVC is used as a film, sometimes in combination with PVF or other topcoats. Polyester also has been manufactured as a film, as have polyurethane and fluoropolymer films such as ethylene-tetrafluoroethylene polymer (ETFE). The following paragraphs describe performance parameters of films.

Mechanical Properties

While the mechanical properties of films vary, discussion herein is focused on ETFE. It is perhaps the most commonly used film, but its use is limited by its mechanical properties. The tensile strength of a typical ETFE film of 0.1-mm thickness is about 225 N/5 cm, less than 5 percent that of heavyweight polyester and fiberglass fabrics. Furthermore, while fiberglass fabrics hold their shape following initial "elastic" elongation under prestress, ETFE continues to stretch or "creep" under sustained prestress or other loads. An ETFE film used in a conventional prestressed tension structure, therefore, tends to lose its prestress over time, and its practical application is generally limited to air-supported or air-inflated roofs, as moderate creep does not significantly affect prestress applied through air pressure.

ETFE films generally yield at elongation of approximately 3 percent, then continue to carry load until rupturing after elongating about 200 percent of their original length. This capability gives films excellent resistance to the propagation of tears (Schwitter 1994).

Scrims using widely spaced fibers of polyester or other material are sometimes sandwiched between layers of film to provide membranes of improved tensile strength. Such composite materials may have excellent tear strength, as well, because the flexibility of the films allows the fibers to bunch at the end of the tear (Geiger 1989).

Durability

While the durability of vinyl and polyester films is limited, ETFE films can be expected to endure normal sunlight and other environmental conditions for more than 15 years.

Light Transmission

Films can be doctored to have little or no translucency but typically are formulated to have a transparency that makes them highly favorable for greenhouse application. The light transmission characteristics of ETFE films over different wavelengths further contribute to their effectiveness in greenhousing. Like window glass, ETFE transmits nearly all of the visible light that strikes it. While window glass blocks a high percentage of UV radiation, however, ETFE is nearly transparent to UV, and the bactericidal effect of UV makes the plants beneath it less susceptible to disease.

Fire Resistance

Fire performance varies among films, and designers are advised to contact manufacturers for fire test data on particular products.

Cost

The material cost of ETFE films is generally in the range of $15 to $25/m^2. However, variations in structure configuration and number of layers of film used in insulated structures make the final installed cost too variable to warrant generalization.

Meshes, Knits, and Other Alternative Materials

There are many variations on the common fabric formulations described in the preceding sections, including those that employ more exotic materials, such as PTFE or Kevlar®, and those with mesh, knit, or other less conventional construction. There also are meshes made of steel wire that are not fabrics at all but are used in

The Teflon® scrim used in the Prophet's Mosque in Medina, Saudi Arabia, allows the canopies to be repeatedly opened and retracted without damage. The decorative seaming gives the canopies the visual quality of open flower blossoms.

Source: SL-Rasch; used with permission.

building construction and share fabric's membrane behavior; therefore, they deserve mention here.

W.L. Gore, the creator of Gore-Tex® fabric, the waterproof material now ubiquitous in outdoor clothing, markets structural fabrics that employ woven PTFE scrims. The material was developed in order to marry the longevity and cleanability of PTFE-coated fiberglass with the resistance to damage from folding and the consequent applicability to seasonal or deployable structures that is provided by PVC-coated polyester. This flexibility and resistance to folding damage also help to reduce costs by eliminating the additional care in handling and packaging associated with shipping and erecting PTFE-coated fiberglass fabrics.

These properties made it ideal for use in the lovely retractable umbrellas covering the Prophet's Mosque in Medina, Saudi Arabia (shown above). The material used in this structure is uncoated and reliant on sewn seams, and therefore has limited water resistance where it is not sloped to provide fast runoff. More recently developed material, used on an entrance canopy engineered by my office for Gore's own facility in Flagstaff, Arizona, is PTFE-coated and thus enjoys the same ability to be heat sealed and watertight as PTFE-coated fiberglass, while retaining the flexibility of the original product.

The material is subject to creep at higher stress levels, and prestress and other sustained stress should therefore be limited to 10 percent of tensile strength. It is currently available in constructions of 300 to 500 dN/5 cm tensile strength. The fabric is white and has

Light polyester knits may elongate 20 percent or more under load, making them appropriate only for shade or decorative applications.

high translucence (30–37 percent) relative to other structural fabrics, while retaining their low absorption and high reflectivity of light.

Kevlar® is a newer material used to make scrims that have very high tensile strength. Because of its poor UV radiation resistance, it requires a carbon black or similar protective coat. Kevlar® has rarely been used in architectural applications to date.

Meshes are fabrics that typically employ a conventional plain weave but use large and widely spaced fiber bundles that leave a grid of gaps in the coated material through which light and water may pass. These materials are useful in shade or decorative canopies where high light transmission or variations in visual texture are desirable. Several firms market such products using PVC-coated polyester materials. Their mechanical properties, durability, fire resistance, and cost do not vary dramatically from that of their conventionally constructed cousins.

Knitting is a fabrication process in which needles are used to form a series of interlocking loops from one or more yarns (shown above) (Kadolph and Langford 1998). Because a great deal of stretch can occur before the looped yarns straighten out, knits have very low stiffness over a large range of strain. This property makes knits easy to form into sharply curved shapes with little or no patterning, but it also makes them unable to carry significant load without large deflection. When stretched out under tension and released, knits may take considerable time to regain their original shape and often undergo permanent creep. The open texture of knits makes it impractical to coat them for watertightness. Because of these characteristics, knits are generally restricted to use in decorative or shade canopies of moderate size.

Wire meshes may be used in applications such as aviaries, where openness to light and air is required in combination with the toughness of steel. Wire meshes include both grids of closely spaced interlocked wire rope similar to cable nets and heavy wires interwoven in the manner of chain link fence.

Cables and Fittings

Cables are key tension-carrying elements in fabric structures, critical to their stability, shape, and load resistance. Cables, like the fabric itself, are selected for their strength and durability characteristics. Their flexibility (in accommodating required bending radii) and appearance also play a role in the selection of cable materials.

The most commonly selected cables use high-strength steel wire rope, in which individual wires are wound into strands that are in turn wound into completed ropes. In typical configurations, 19 or 37 wires are wound into each of the seven strands that form a cable. Wire rope combines high strength (ultimate tensile strength of up to 1,900 Mpa) with relative flexibility and ease of handling. At connection points, the ropes can be shaped to radii of about 15 times cable diameter with no reduction in allowable stress (ASCE 1997). Their flexibility adapts to the tightest radii required at catenary, valley, ridge, or radial applications and accommodates reasonably sized saddles where the cable must bend around steel fittings. Wire rope can be rolled to package readily for shipment to the site, and it can be manipulated into position with relative ease in the field.

Where less flexibility is required, designers may select steel structural strand, in which larger diameter wires are wound into a single strand that forms the finished cable. In general, strands have slightly higher tensile strength and stiffness than wire rope of the same diameter, but they can only be bent to a minimum radius of about 20 times cable diameter without weakening. Their characteristics are well suited to uncurved guy or tie-back cables.

Tensioned fabric structures also provide limited applications for steel tensile rods, formed of a single round or rectangular cross section without individual wires. The use of tension rods is limited to applications where no curvature is required and where the inflexible length of the rod does not present an impediment to shipping and handling.

On smaller structures using vinyl-coated polyester fabrics, particularly those for temporary or demountable applications, polyester webbing may be used in lieu of steel cables both to reduce initial cost and to provide excellent flexibility for tight radii and ease of handling. The limited strength and stiffness of webbing restricts its use to applications with working loads of 80 kN or less, using currently available products, and assuming a webbing width of 100 cm.

Durability is a critical consideration in cable material selection. Steel cables, particularly in exterior applications, must have adequate protection from corrosion. This protection is critical to the durability of the cables, but it also serves to prevent staining of adjoining fabric material. This is most often provided through the use of galvanized wire. If manufactured to ASTM standards, ropes with "Class A" zinc coating on all wires are generally suitable for interior and most exterior applications, while severe exposures may employ Class B or C coatings on outside wires.

Stainless steel cable, although it has higher initial cost, may be used where a higher level of corrosion protection is required, or where the clean and high-tech look of the material's bright silvery finish is desired.

The use of webbing eliminates the corrosion danger associated with steel cable. Durability is limited by the effects of UV radiation on the polyester material, however, generally limiting its effective life span to 7 to 10 years.

Cable fittings should be of a type as required to suit project detailing requirements (see Chapter 6), and fittings may be stainless steel, galvanized, or painted, as required to suit aesthetic, budgetary, and durability requirements.

Supporting Elements

Fabric roofs, like all tension-based structural elements, require compression and bending elements to support them and bring their forces down to grade. Supports often provide, as well, a means of applying tension to the fabric. Reinforced concrete, wood, and metals all have the ability to provide tension and bending resistance economically, and each has found applications in fabric roof support (see the photographs on page 84).

However, the majority of fabric roofs rely for their support on structural steel, a material that is readily and economically available in cross sections of high bending and compression resistance and that is readily transportable and easily erected in the form of overhead arches or tall ground-supported masts. Another factor weighing in favor of the use of steel is the precision with which it can be shop fabricated and therefore interfaced accurately with a carefully patterned and fabricated fabric membrane. Steel provides, as well, an appropriate visual marriage to fabric, with smoothly textured surfaces and slender cross sections that use the material's high strength to advantage.

One of the defining characteristics of tensioned fabric roofs is the number and complexity of their connections: fabric, multiple cables, and supporting elements are often joined at a single point. Steel, as a material that can readily be formed in varying cross sec-

The Buckingham Palace Ticket Office uses straight and curved timber elements to gracefully support the roof canopy of this seasonally erected structure.

tions, or cast, cut, bent, rolled, punched, welded, or bolted, is above all else a material that facilitates the complex demands of connections. It may be this feature that provides the most compelling rationale for the merging of steel with fabric.

The nature of tensioned fabric roofs does not generally create a demand for steels other than the mild carbon steel most common to construction. Finishing of the material can be critical to its success, however, particularly in applications where the supporting structure is exposed to the weather or in unusually damp interior environments. Galvanizing generally does not provide an appropriate appearance, so painting must be done using materials and processes that ensure adequate corrosion resistance, and field workers must use special care in avoiding impact or other damage to finishes. Where steel interfaces with other materials, such as aluminum clamp bars, special care must be taken to isolate the two materials against the effects of bi-metallic corrosion. In addition to potential long-term effects on the strength of the structure, corrosion presents the danger of unsightly staining on light-colored fabrics and must be carefully guarded against in design.

5 Form Finding and Analysis

Contemporary tensioned fabric roof structures are successors to a tradition of tent building that dates back to prehistoric times and includes more recent tents for circus and military applications. Emerging technologies, however, make these contemporary structures fundamentally different from those created as recently as 30 years ago. Newer materials ensure fabric life spans in excess of 20 years, even in harsh environments, while meeting strict fire safety standards. High light reflectivity and natural daylighting provide energy efficiency in warm and bright climates, while evolutions in translucent and flexible insulation provide the opportunity for improved energy performance in cold climates, as well. Recent designs have provided R values of 12 or greater while maintaining sufficient translucency to eliminate artificial lighting under full daylight conditions. These developments have given tensioned fabric roofs application in permanent, high-occupancy, enclosed structures, and in buildings where energy-efficient, conditioned spaces are required.

While materials advances have improved many measures of fabric roof performance, it is newer structural design methodologies that have been instrumental in the evolution of traditional tents into high-grade works of architecture. With these methodologies, a construction genre that was once restricted to conventional tent shapes and hampered by wrinkled forms and unreliable behavior has evolved to one of porcelain smooth surfaces of wildly divergent shape with predictable performance under all environmental conditions.

This chapter provides an overview of the shaping and analysis of tensioned fabric membranes using methods that include hand computation and physical and computer modeling. The mathematical complexities of these methods are of limited interest to most readers and are well documented elsewhere; therefore, they are not included here.

By measuring the forms of soap films stretched between fixed boundaries, designers learned to model the shapes of uniformly tensioned membranes.

Source: Larry Medlin; used with permission.

Non-numerical Shaping Techniques

Up to the 1960s, tent shapes emerged from the gradual evolution of previously realized forms, with model studies and simple hand calculations used to evaluate shapes and structural adequacy. The German designer Frei Otto developed the use of physical modeling in membrane roof design. His best-known models were those made using soap bubbles (shown above). Soap film models have several defining properties. First, in stretching themselves between any rigid edges or supports (analogous to the cables, arches, or masts in a roof), they always form the shape having the least potential energy. In essence, the soap film relaxes, finding the shape that minimizes the overall tension in the soap membrane. These "minimum-energy" surfaces have two other interesting properties: (1) they provide tension stresses in the surface that are equal at all points and in all directions, and (2) they have the minimum surface area that will join the film to its edges and supports. By stretching thin soap films between any desired edge condition and graphically recording the resulting shapes, Otto found that he could replicate the shape that a fabric membrane or cable network would take in order to be stable and uniformly prestressed over any proposed system of mast or arch supports.

Soap film models continue to provide a valuable tool for conceiving and visualizing tension structure forms, but their usefulness in design is limited by the difficulty of measuring their shapes accurately. Techniques also were developed using heat-shrink polyvinylchloride (PVC) foil that cools to a fairly rigid model form from which paper-cutting patterns can be taken. Otto developed the technique of using fine wire and small cable clamps to develop tension structure shapes. The latter method was used to obtain cable net and cladding

The innovations of Frei Otto's West German Pavilion for Expo '67 were made possible, in part, by the use of wire models with suspended loads.

Source: Larry Medlin; used with permission.

[LEFT] Exterior view.

[ABOVE] Wire model.

geometry, forces, and deflections for both the West German Pavilion (shown above) and for the swimming pool roof at the Munich Olympic Park (Liddell 1989).

Elementary Analytical Procedures

Analysis of the forms developed through wire models or other techniques was carried out using simple hand calculation on two-dimensional forms. These calculation techniques also illustrate the basic principles of fabric membrane load carrying that form the basis of the complex computer software used in contemporary design.

Under vertical loads, a cable or other tension element assumes a catenary shape akin to that of a suspension bridge cable (see [a] in the drawing on the next page). When these loads are uniform, like most design live loads, the catenary shape is coincident with a parabola, and the reactions at the ends of the cable are calculated according to the following formulas:

$$V = \frac{wl}{2}$$

and

$$H = \frac{wl^2}{8h}$$

The resultant force, F, in the cable is given by the formula

$$F = (V^2 + H^2)^{1/2}$$

For cables loaded perpendicular to the line joining their end points and with sag ratios (h/l) of 0.2 or less, the tension force may be approximated by the value of H.

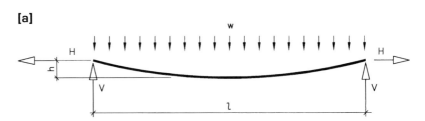

A cable or fabric membrane will assume (a) a catenary curve under vertical loads (dead or live load) and (b) a circular curve under loads perpendicular to the surface (wind).

Under loads perpendicular to the cable, such as wind, the element assumes a circular curve (see [b] in the drawing above). The cable tension is calculated according to the formula

$$F = \frac{w(l^2 + 4h)}{8h}$$

where w is the applied force per unit of length along the circular arc.

Returning to the parabolic element, the original arc length of the cable can be approximated by the formula

$$A = 2\left[\frac{l^2}{4} + \frac{4}{3}h^2\right]^{1/2}$$

and the change in its arc length under load (δA) can be calculated by the formula

$$\delta A = \frac{FA}{E}$$

where E is the modulus of elasticity of the cable, calculated as the ratio of stress to strain (σ/ε) under load. Modulus of elasticity is the property most commonly used to define the stiffness of materials of all types.

The new arc length, then, is

$$A' = A + \delta A$$

Transposing the above formula for A, the sag of the membrane can be recalculated by the formula

$$h' = \frac{\sqrt{3}}{4}\sqrt{A'^2 - l^2}$$

88 THE TENSIONED FABRIC ROOF

The revised sag is then input to the original formula for F to compute a revised tension force. Because of the increase in sag due to the change in arc length, the tension force will be less than computed originally, and all of the formulas can be cycled through in iterative fashion until the final geometry and forces are approximated. The changes in geometry that occur under load can have a dramatic effect on the final stresses in a cable, and calculation of these nonlinear effects is an important part of the analysis process. If a fabric membrane can be viewed as a network of intersecting cables, we see that the deformation of a fabric roof under load has a significant impact on its internal stresses.

Prior to the availability of membrane analysis software, the simple analytical techniques described above were applied using geometry measured from physical models. The computer automates what is a laborious process of hand calculation, and accurately models the effects of three-dimensional behavior, which is far more complex than the two-dimensional model developed above.

A simple example illustrates the methodology and the effect of geometry changes on internal forces. Consider a parabolically curved membrane segment with the following input parameters:

l = 10.0 m
h = 0.5 m
w = 0.8 kN/m
E = 1,200 kN/m

The initial iteration yields the following results:

F = 20 kN
A = 10.067 m
δA = 0.168 m
A' = 10.235 m
h' = 0.944 m

Following through with successive iterations, the shape converges to one having a rise (h') of 0.805 meters and a tension force (F') of 12.4 kN. In the example, then, using span-to-sag ratios, loads, and stiffness that are realistic for a tensioned fabric structure, consideration of nonlinear effects changes the resultant tension in the membrane from 20 kN to 12.4 kN, a 38-percent reduction.

Computerized Techniques

The Munich Olympic Park design represented the pinnacle of tension structure design based on physical modeling. It also provided a bridge to the modern era of computerized shaping, analysis, and patterning. As the elongation under prestress of a 25-m-long cable may be only 15 mm, a length error of 5 mm (very reasonable for geometry derived from a carefully constructed wire model) results in a prestress error of more than 30 percent. Noting this, the Olympic Park engineers began to seek greater accuracy through purely analytical solutions, and they developed software in time to be successfully used in the design of the Sports Hall portion of the Munich complex (Holgate 1997).

Computers had been used for the analysis of fabric roofs before this pioneering German work, but the engineers using them generally relied on adaptations of general-purpose structural analysis software that did not address certain key aspects of fabric structures. Most important, they assumed that the shape of the fabric was the same before and after loading, when actual deformations were often large. The new software, though, had the ability to apply loads in small increments and recompute the structure's shape after each load application to provide an accurate determination of the final shape and stresses.

While the Munich structures had acrylic roof surfaces supported on steel cable networks, the technology was adopted and further developed by engineers in the Geiger Berger office in New York City for use in designing fabric membranes without cable nets. They learned that, by modeling a strip of fabric as if it were a cable having equivalent axial stiffness, a good approximation of the proper shaping and stress under load of the fabric membrane could be realized.

By the mid-1970s, these computer programs were as amazing for their relative accuracy and analytical power as they were stupefying in their cumbersome methods of data input and interpretation. The programs were used in the late 1970s to design the Hajj Terminal, soon to become the largest fabric roof in the world; and analytical results about the forces in various members in the structure were validated within a few percent by test results on a prototype constructed later. As part of the team of engineers preparing that analysis, however, I spent many sleepless nights preparing input data for computer runs that cost $5,000 to $10,000 each on the first Cray supercomputers. The programs were limited in their means for automating the generation of geometric data, and each of several thousand "nodes" defining the shape of the fabric membrane was defined and input by hand. The graphic interface to plot this input geometry and assist the engineers in verifying its numerical accuracy was primitive.

The refinement of computing technology increased rapidly thereafter, however, until by the mid-1980s, engineers had automated the input of most data and could instantly generate plots that showed the structure's geometry and deformation under load. Engineers at Birdair, America's largest fabric structure contractor, could sit at the computer with the owner or architect of a proposed building and "walk him around" the structure—observing an isometric plot of the

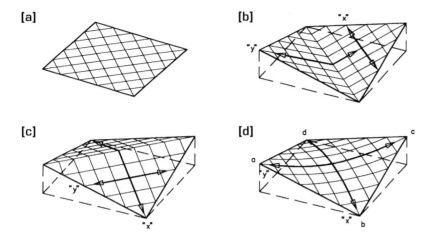

A cable net modeled by the force density method yields the characteristic anticlastic tension structure form (d).

shape from any location, even moving inside the building in order to get a sense of its interior volume (Huntington 1994).

There are several methodologies for finding or modifying shapes on the computer in common use today. One, derived from the original German method used at Munich and elsewhere, determines minimum energy/minimum surface shapes analogous to the Otto soap films. The technique belies its roots in cable net structures by modeling the fabric as a gridwork of cables, each representing the stiffness of a narrow strip of fabric (shown above). If such a cable network is used to represent the membrane inside a square, flat perimeter, all of the cables will run in straight lines from side to side (see [a] above). A fabric roof that is flat like this will be subject to large deflections under load, like a trampoline, unless it is highly tensioned in the manner of a drumhead.

If two corners of the perimeter are lifted so that the edge no longer lies in a plane, it is no longer possible for all of the cables to pass from side to side in straight lines. Two extremes of shape are possible with the revised support configuration shown. In the first (see [b] above), cables in direction "x" remain straight, as if they were of infinite stiffness, while those in direction "y" kink at mid-length to accommodate the imposed geometry of the perimeter supports and "x" cables. In the second (see [c] above), the "y" cables remain straight, as if *their* stiffness were infinite. The real structure will find a shape intermediate between these two extremes (see [d] above), with both "x" and "y" cables deformed such that there is a constant "density" of force (the ratio between the load in a member and its length) throughout the structure. The elegantly curved shape that results from this "force density" method is that which stores the minimum total potential energy in the cables and has the minimum total cable length. The form is analogous to one of Otto's minimum-energy/minimum-surface-area soap bubbles. The numerical force density technique has several advantages over the soap bubble, however, including the automated and accurate documentation of geometry and the fact that the force density can be made different about two opposing axes when the designer wishes to manipulate the shape or to vary prestress levels in the two directions. Assuming that the force density "cables" are spaced closely enough, they model the shape of the actual fabric roof with sufficient accuracy, and analysis of the fabric structure under load can follow.

Membranes of fabric or other material characteristically undergo large deformations under load, and accurate analysis generally requires iteration: analysis of fabric stresses under load using the initial shape, calculation of deformation and an adjusted shape under the calculated stresses, reanalysis using the new shape, recalculation of deformation, etc. Contemporary analysis has automated this process, allowing rapid iteration until the shape has "converged" to a geometry that approaches equilibrium under the applied load.

While contemporary structural analysis software is good at modeling the geometric nonlinearity discussed above, it generally does not have the ability to model the material nonlinearity associated with crimp interchange or other characteristics of the material under load. Engineers instead select single values for modulus of elasticity (defined as the ratio between stress and strain, where stress in this instance has dimensions of force per unit width of material) for the material in the warp and fill directions. The appropriate values will represent an average of the stiffness that occurs over the normal range of use.

Design Evaluation and Refinement

Analysis software output typically includes a listing of membrane tension stresses in warp and fill directions in all areas of the membrane and under all selected loadings. The engineer compares these stresses to the allowables for the fabric selected, usually given as the strip tensile strength of the fabric divided by a factor of safety. Strip tensile strengths for various fabrics are typically provided by manufacturers, and tests can be readily and inexpensively performed on new products, as required.

Appropriate factors of safety, however, are not so readily derived. At present, it is common practice to use factors of safety against strip tensile strength of 4 for short-term load cases that include wind and 5 for live load cases, using forces calculated by working stress procedures. Note that design live load is used to account for such things as snow or workmen or other nonpermanent loads on top of the roof, and that all stress calculations must account for the effects of prestress and the structure's own dead load. This simple approach, however, ignores a number of factors that influence the strength and reliability of the roof: degradation of the fabric over time, damage due to handling, material tear strength, reliability of seaming, and the level of stability and redundancy of the supporting structure in the event of damage to the membrane. Standards currently under development by the American Society of Civil Engineers (ASCE) are expected to include these factors in their methodology for derivation of more rational factors of safety for specific applications.

Often, initial analysis will indicate overstress in one or more areas under one or more load cases. Several strategies are available to the designer for alleviating overstresses, as follows:

1. **Alter the initial prestress shape.** In most designs, initial prestress provides a uniform stress in all areas of the fabric membrane and about both axes (in the manner of a soap film). Under load, however, overstress may occur about a single axis over some area of the membrane. By reducing the prestress in the overstressed direction relative to the other, curvature in the initially overstressed direction is increased so that it resists applied load with less increase in stress, thereby eliminating the overstress. An example is shown in [d] on page 91. Live loads (acting downward) might create an overstress in the fibers oriented in the "y" direction. By reducing the prestress in these fibers relative to the "x" fibers, the initial equilibrium shape will be depressed downward to provide greater curvature in the "y" fibers. Under a given live load, the stress increase in the overstressed "y" fibers will be less, so that the overstress is eliminated. In the example, the increase in curvature of "y" fibers is associated with a decrease in curvature of the "x" fibers and consequent increase in "x" fiber stresses under upward loads such as wind. The designer must beware that solving one problem does not create a new one!

2. **Change support conditions.** Increasing fabric curvature in order to reduce stress under load also can be accomplished by altering the geometry of the supports. Referring to drawing [d] on page 91 again, "y" fiber curvature might be increased by raising corners "a" and "c" or lowering corners "b" and "d." Unlike the simple alteration of prestress shape, however, this change in support geometry increases the curvature of the "x" fibers, as well, thereby avoiding the possible introduction of new overstresses. The change in geometry is not without possible cost, however. It may change the appearance of the structure significantly, and the increase in overall height of the structure may have such side effects as increased mast or wall height (resulting in higher wind loads and increased fascia cost) and greater enclosed volume (resulting in increased energy usage).

3. **Add cables.** By adding cables in line with overstressed fabric fibers, some of the load initially carried by the fabric is transferred into the cables. Given the high stiffness (as measured by modulus of elasticity) of steel cabling, fabric overstresses can be eliminated rather effectively in this manner, though the cable will of course increase the cost of the structure.
4. **Change the fabric.** Changing the selection of fabric to one having higher allowable stress provides the simplest solution to the problem of fabric overstress. Where overstresses are about one axis only and over only a limited area, however, changing the fabric is also the least elegant of the possible solutions because it imposes higher materials cost on the entire structure when the stronger material is required over only a portion of the area.

Patterning is the last stage of the computational process in which the geometry determined during shape finding and validated under load in analysis is converted into actual cutting patterns for the fabric. Patterning might appropriately be considered the first step in fabrication rather than the last step in design, similar to the preparation of shop drawings in steel construction. Because of this, patterning is addressed in detail in Chapter 7, Fabrication and Erection.

Loading

The engineer analyzes the fabric roof under a variety of wind and live load cases, each in combination with the initial prestress and dead load. At present, loading requirements for tensioned fabric roofs are generally governed by building code requirements that were drafted to address structures that have neither the flexibility nor the curved form of fabric, although standards under development by ASCE are being formulated to address this deficiency. The load-bearing characteristics of tensioned fabric structures are governed by the high deformability of membranes under load, and may be generalized as follows:

1. Dead load from the membrane and cabling is usually less than 50 N/m^2 and has negligible design impact.
2. Point loads such as heavy lights, signs, or scoreboards present special design problems due to the high deformability of membranes. Heavy loads must generally be supported from rigid mast or arch supports or at angle changes in cabling, and care must be taken that they do not induce low points in the field of the fabric that will result in ponding of water or snow.
3. Building code roof live loads were established to account for construction phase loads such as roofing materials that are not relevant to fabric construction. Code provisions generally make no loading exceptions for membrane construction, however, and

fabric roofs are therefore typically designed for normal live loads, subject to code provisions for live load reduction based on tributary area. In the model codes of the United States, this reduced live load has a magnitude of 575 N/m^2. Obvious exceptions have been made for air-supported stadium roof membranes, which are typically held up by internal air pressure of approximately 250 N/m^2 and therefore have much lower live load capacity. Because wind loads are usually of equal or greater magnitude, however, the common over-conservatism in live loading usually does not impact the maximum stress in the fabric and hence fabric selection, although it may result in increased loading on some cables and supporting structural elements and foundations that resist downward loads.

4. Earthquake loads are generally not a factor in design, even in areas of high seismicity, because of the low mass of the fabric roofs.

5. Wind is often the predominant loading on the fabric roof. Depending on the slope of the roof and its windward or leeward exposure, wind pressures may vary in magnitude across the roof and act either in or out on the fabric surface. As wind most often creates outward suction on the membrane, however, it generally acts in opposition to live or snow loads, often resulting in uplift forces on foundations and causing governing forces in certain cable elements. The membrane must have adequate curvature and pretensioning to resist wind loads without large areas of the membrane going slack and falling vulnerable to damage from excessive flutter. Building codes and other reference materials provide only limited guidance to the determination of wind load on roofs of complex shape. Where a roof is large or has low stiffness due to minimal curvature or prestress, it may have a low natural frequency of vibration such that dynamic effects may significantly increase wind load magnitude. For structures with a natural frequency much longer than 1 second or with unusually complex shapes or surrounding site conditions that make wind loading unpredictable, wind tunnel testing may be warranted.

6. Because roofs must be configured to avoid ponding of water, rain loads generally are not considered in the analysis.

7. In colder climates, snow loads may govern the design of both the membrane and some cables and supporting members. Appropriate consideration must be given to concentration of loads due to snow drift. In major snow country structures, model testing may be appropriate to establish design snow loads.

8. Temperature variations in fabric roofs have not been known to cause problems in service and are generally not considered in design.

Both sails and fabric roofs must be held taut over supporting structures in order to avoid wind flutter.

The Fruits of Accurate Shaping and Stressing

Together, shaping, analysis, and patterning programs form the triumvirate of computational tools that have made predictable performance of widely varying fabric roof forms a reality.

Like a flag or sail, a fabric roof is unstable unless it is stretched taut over some supporting framework. With the often complex curvature of these roofs, patterning the fabric into a shape such that a fairly uniform level of pretensioning exists over the entire surface area of the roof can be a daunting task. Reliable structural behavior, though, in which some fibers do not become overly tensioned while others flutter excessively due to a lack of tension, is dependent on providing a regulated level of prestress throughout the surface of the fabric.

The developments in computing technology that have taken place since the 1950s, discussed in the previous sections, have dramatically improved the accuracy with which the shape of the finished structure and the stresses in fabric, cabling, and supporting elements can be predicted under load. Accurately regulated levels of prestress can now be reliably provided, even on structures of unusual shape, and the resulting improvements in structural behavior have been enormous. On an old circus tent or other fabric structure that preceded the new computing technology, wrinkles (representing a lack of prestress at some location on the structure) and patches (often representing a location where the fabric tore to relieve a local overstress) were ubiquitous. A few years ago, though, Lee Erdman, then President of Birdair, was able to boast of a corporate policy of "no wrinkles, no patches," and usually make good on it (Huntington 1993). Determination of appropriate loading and accurate analysis have benefits that go beyond aesthetics or corporate pride, however, as they are requisite to preventing overstresses in the fabric that may result in fabric tears or other damage.

Accurately regulated prestress has brought enormous improvement in the visual sophistication of tensioned fabric roofs; so too has the lessening of the visual clutter of turnbuckles and other devices used to allow adjustment of membrane shape and prestress in the field. With contemporary analytical techniques offering a high likelihood of accurate shaping and prestress, designers often prefer to omit devices that allow field adjustment of cable length or other geometries of the structure, having observed various incidences in which well-intentioned field personnel created unintended overstresses or tears in the course of their efforts to perfect the tensioning in a structure (Huntington 1994).

With the exception of air-supported and air-inflated structures, tensioned fabric roofs rely on anticlastic surface shapes to provide stability under load. Saddles, hyperbolic paraboloids, and the char-

The porcelain smooth fabric of the Pavilions at Buckland Hills in Manchester, Connecticut, is emblematic of the capabilities of contemporary computerized shape-finding, analysis, and patterning.

Source: Birdair, Inc.; used with permission.

acteristic cone shape of traditional tents all have this property. With their opposing curvatures, loading inward or outward against the fabric surface causes an increase in fabric stress about one axis and a decrease about the other. If sufficient prestress and curvature is provided, large areas of the fabric are never allowed to go slack in a properly designed roof.

During erection of a fabric roof structure, however, the membrane may lie draped over the supporting structure without any prestress, or a portion of the roof may be tensioned while another is not. Stable shapes with appropriate anticlastic surfaces may not occur throughout erection, and large secondary stresses may be introduced by the imbalance of tension throughout the membrane. This may be of little importance in roofs of moderate size and simple configuration, where a single crane can lift the entire fabric membrane into place over a supporting framework and prestress it within a matter of hours. In larger contemporary structures like the cable domes, however, it may require weeks to stress the fabric into its final, stable position, during which time the loose fabric is subject to self-destruction from wind flutter.

Several years ago, I participated in the investigation of the failure of a fabric canopy more than 150 m in length that was draped loosely over a Hollywood stage set without pretensioning. Moderate winds along the length of the structure created waves in the canopy that rolled repeatedly from one end of the structure to the other until the fabric tattered like an old flag. Computer analytical techniques have made such disasters avoidable and have created not only the ability to evaluate the stability of a structure throughout the different stages of erection and tensioning of the fabric, but also to accurately analyze the secondary stresses that may be introduced by the adjustable cabling or telescoping masts often used to tension contemporary fabric roofs.

Enriching the Vocabulary of Forms

In addition to improving the accuracy of shaping and stress analysis, the new computing technology has created the ability to confirm the viability of unusual shapes and to predict their behavior under load—an enormous boon to designers seeking to make a bold statement or respond to unusual site parameters. Walter Bird, the founder of Birdair, Inc., and grandfather to much of the current state of the art, notes that "before the computer, we were pretty well limited to geometric shapes that we could calculate by hand." The high-profile air-supported spherical Radome structures (shown on page 43) that he built beginning in 1948 were spherical forms adapted to the limits of hand analysis techniques. Repeated construction of similar structures allowed gradual refinement in an iterative approach to a nearly perfect design. "When we got the computers," Bird says, "we could do in a few seconds what it took weeks to do before," and the evolution of radical new shapes began to take hold (Bird 1999, personal communication).

Prior to the development of appropriate analytical techniques, the development of structural forms (using fabric, concrete, metal, timber, or any other material) has been not unlike the evolutionary process discovered by Darwin. Successful innovations in form were carried forward to the next generation of structures, while failures were not reproduced. Gothic cathedral form followed such an evolution, although the mass of stone used and the difficulty in cutting it meant that evolution was measured over the course of centuries, not years. Working without mathematical tools, the medieval masons strove to make each cathedral a little taller and more slender than the last, carrying successful experiments forward to the next generation and abandoning those design changes that led to falling stone.

Contemporary engineers not only have the ability to consider variations in mast location, arch curvature, scale, and other refinements to a design, but also they have the capacity for developing reliable, large-scale roofs using structural systems with little or no precedent. Atlanta's Georgia Dome (shown on page 52) offers an example. An oval-shaped roof with a maximum span of 235 m, it draws inspiration from both the tensegrity concept of Buckminster Fuller and the cable domes created by David Geiger. None of Fuller's domes were built, however, and the radial cable system created by Geiger was substantially different in both shape and structural behavior from the triangulated system created by engineer Weidlinger Associates for Georgia. Yet the analytical tools available to Weidlinger and contractor Birdair allowed them to design and erect one of the world's longest span roofs without the opportunity to build confidence working on smaller roofs of similar design.

6 Connections and Detailing

Material selection, the subject of Chapter 4, is critical to the durability, energy efficiency, and fire safety of fabric structures, while the analytical techniques discussed in Chapter 5 give the roofs their precise shaping and ability to resist extreme winds and other loading. This chapter addresses a third aspect of design that is seldom addressed in the tension structures literature: the structural details that fasten the elements of the roof together and provide a means for tensioning the fabric. Details are critical to the performance of any type of structure. In fabric roofs, though, connection details are generally left exposed in the finished structure and are critical to their appearance as well. The elegance and expressiveness of connection details is inextricably linked to the aesthetic success of fabric architecture and necessitates unusually close collaboration between architectural and structural designers.

Every structural material has both a vocabulary of construction elements and a finite number of commonly recurring connection and detailing problems. Steel structures, for example, are typically constructed with beams, columns, braces, and diaphragms. Their design requires, among other things, the development of beam-to-beam, beam-to-column, column-to-column, column-to-brace, and column-to-foundation connections. The performance of these connections has been well researched, and procedures for typical designs have been incorporated into the codes of practice.

Fabric structures also have a vocabulary of elements: fabric, cables, masts, arches, and beams or other supporting elements. The design of a fabric roof requires resolution of the connections between these elements. The solution of connection problems may be more difficult than it is with steel or other "conventional" structural materials, however. First of all, while framing with most conventional struc-

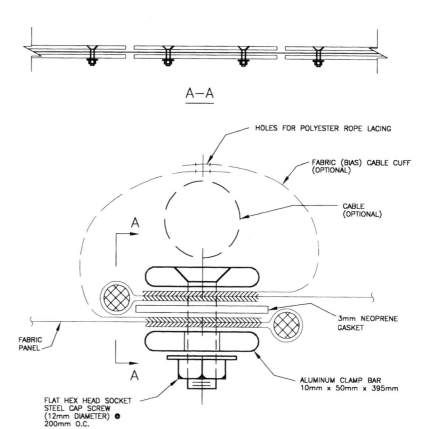

Fabric field joints can be made by sandwiching two fabric roped edges (see Section B-B of the drawing in the middle of page 140) between clamp bars. The joints are made to align with a cable where possible, as shown here. Typically, the ends of upper and lower clamp bars are aligned, as shown in Section A, rather than offset, so that continuously linked clamped bars do not acquire stiffness similar to a cable, attracting forces for which they were not designed.

tural materials is rectilinear, with members meeting at right angles, the angular geometry of fabric structures is almost always irregular. The fabric designer must solve detailing problems in a manner that is economical and, because connections are generally left exposed, visually elegant. He must provide for substantial movement and weather exposure in service and, at the present time, must develop these details with little aid from codes of practice or textbooks. This chapter provides the conceptual framework for the designer of the various characteristic connections in a tensioned fabric roof.

Most connection design situations provide for more than one technically acceptable solution, and the most successful structures employ connections designed to provide simple and direct load paths and a visual expression of the flow of forces that is intuitively clear to both designer and layman. These structures have a congruity in function and appearance among the various connections and a common vocabulary of design elements that provides clear indication of commonality in function. While the large-diameter catenary cables on a structure may require speltered end fittings, and smaller diameter catenaries might more economically use swaged fittings (see the drawing on page 109), for example, altering fitting type within the structure implies a variation in function that does not exist and is visually dissatisfying.

In the detailing of connections, the designer is advised to follow what may seem the path of *most* resistance: solving the design problems of the most complex or heavily loaded connections first, then

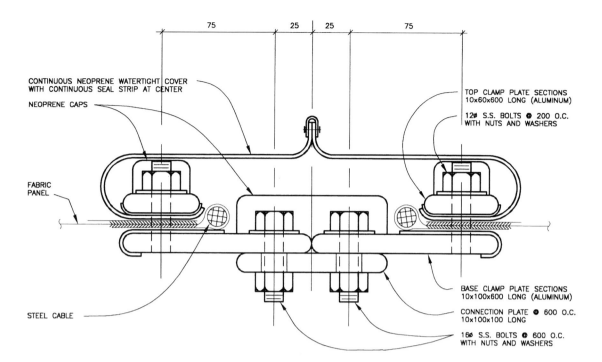

lightening or simplifying them as required elsewhere. "If the most crucial, difficult detail is not burdened with boundary conditions resulting from other details," notes Jörg Schlaich, one of the masters of contemporary structural design, "it has a chance to become simple" (Ito 1991).

Wider base clamp plates allow the fabric to be pulled to position in place.

Fabric Joints and Terminations

Fabric Joints

Each of the characteristic fabric termination types—fabric-to-fabric, fabric-to-cable, and fabric-to-rigid support—has particular design requirements and multiple solutions. The most basic connection of all, the joining of two strips of fabric at a seam, is a shop fabrication process discussed in Chapter 7. Fabric-to-fabric connections are also made in the field when the size of shop fabric assemblies must be reduced to accommodate limits on the size of panel that can be maneuvered within the fabrication shop, shipped, or handled during erection.

Depending on the logistics of erection, field seams may be made either on the ground or in-place atop the roof's supporting structure. Roped fabric edges secured between clamp plates are used most often (see the drawing on previous page), while other, typically smaller roofs utilize lacing or other simple joining mechanisms between panels. Where the field joint is made in-place and substantial tension must be applied in order to draw the fabric joint together, wider base plates may be substituted for conventional clamp bars so that come-alongs may be attached to draw the fabric together (see the drawing above).

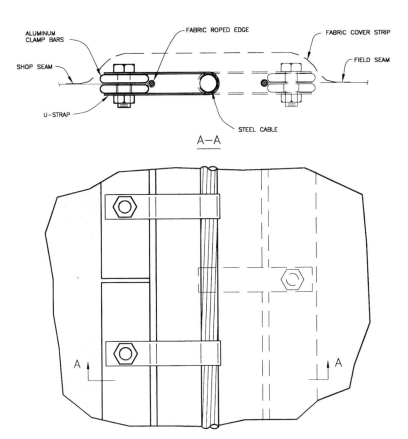

U-strapping clamped fabric to cables is a means of providing both fabric seams and fabric terminations at catenary cables (the undashed portion of the figure). By exposing the cable in the latter configuration, the transfer of tensile forces to the catenary is expressed in a manner not possible with simple cable cuffs.

In many cases, fabric field seams are aligned with cable locations so that the cable rides directly over a standard double clamp joint (see drawing on page 102). The cable may tend to realign itself to one side or the other of the joint in an aesthetically clumsy manner, a problem that can be corrected by periodically substituting an upturned channel for the standard upper clamp plate or by using the more elaborate detail shown in the drawing above.

Termination at Catenaries

Fabric terminations at catenary edge cables are most expeditiously detailed by providing a continuous edge cuff through which the cable rides (see Section C-C of the drawing in the middle of page 140). The cuff material is generally cut at a 45-degree bias to the main fabric panel so it can be curved to fit the line of the cable without wrinkling. Tightly curved cuffs, or those made on stiff fabric, may still wrinkle perpendicular to the cable in an unattractive manner. Closely spaced slits may be made on the inside edge of the cuff to increase its flexibility, like kerf cuts in timber. If the slits are made, another continuous strip of fabric must be sealed over them to provide appropriate reinforcement.

The designer must take care in sizing the cuff so that a clevis jaw or other cable termination will slide through it without binding. These fittings become prohibitively bulky on larger diameter cables so that one end of the cable is terminated in a threaded "stud end" onto which the clevis jaw is threaded after the cable has been run

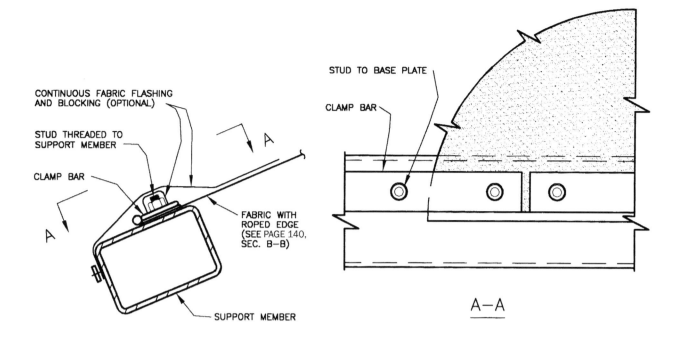

Single plate clamped edges.

through the cuff. Alternatively, the edge of the fabric may be secured between clamp plates that are bolted to U-straps that capture the cable, a single-sided variation of the fabric-to-fabric connector shown on page 104.

Termination at Supporting Structures

The connection of fabric to a rigid edge support of steel, concrete, or other material can be accomplished in simple fashion with single clamp plates, as shown in the drawing above.

Catenary edge terminations are characteristically used around the perimeter of open-sided structures, while rigidly clamped edges are typically used to provide a direct interface between the fabric roof over an enclosed space and the supporting walls or other conventional construction. On some structures, designers wish to maintain the "free" form of the roof by allowing its edge to spill over the top of the wall and terminate at a catenary edge. The interface between the rigid, straight wall and the flexible and curving roof becomes more problematic in such designs. At the El Monte Resort, our design achieved this by adding a return skirt that is sealed to the underside of the roof membrane in line with the ridge cable joining the tops of the low masts. The skirt is then secured to the top of the window wall below (see the photographs on the next page). Others, such as the Denver Airport Terminal, link the underside of the membrane to the top of the wall with continuous inflated fabric tubes that are able to flex upward and downward as the membrane deflects under load.

Corner Terminations

There are special detailing problems associated with the termination of the fabric at "corners," those locations at the edges of the mem-

The El Monte Resort maintains a free catenary edge while providing a weathertight enclosure with a return skirt that is nearly invisible from both interior [ABOVE] and exterior [RIGHT].

Source: Eide Industries; used with permission.

brane where cables terminate at masts or other supporting elements. Small errors in patterning or cable length can have critical effect at these locations, where the fabric necks down to a small width. In addition, tension in the fabric tends to pull it away from the supporting member, causing it to ride up the cables and away from the support. This effect is generally not addressed in currently available software, which assumes that fabric and cable share the same nodal geometry such that sliding between fabric and cable is not modeled. The effect can be pronounced when the angle between the two cables is acute. While reinforcement of the fabric in this area is helpful, some supplemental mechanism for restraining the fabric is generally required. Often, the fabric is terminated at a roped edge, then clamped and held in position with one or more turnbuckles that allow fine-tuning of the final position (see the photograph on the next page).

"Trimming" the fabric with clamp bars and turnbuckles provides a workmanlike method for local restraint and adjustment of the membrane. Such terminations can be visually busy, however, and the use of "hard" clamp bars can pose a threat to "soft" fabric in conditions where the fabric goes through significant geometry changes during erection or in service. Alternatively, the fabric may terminate at a secondary cable spanning between the cables at the edges of the fabric (see the drawing on the next page). The flexibility of the secondary cables makes them forgiving of movement, but the adjustability of turnbuckled terminations is lost. In patterning a membrane where fabric is terminated at a secondary cable, the designer should ensure that the final shaping accounts for the inward displacement of the main cables that results from the tension in the secondary cable.

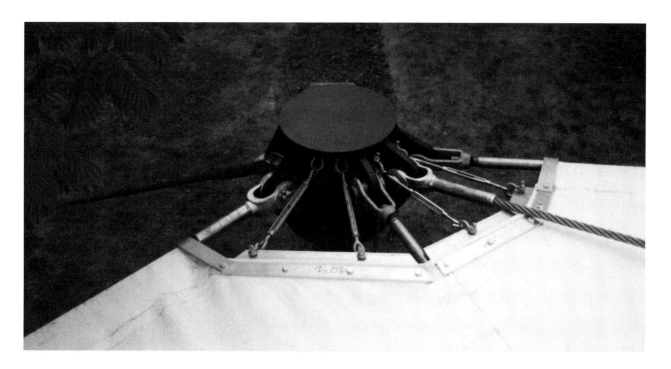

[ABOVE] The turnbuckles in our design for the National Semiconductor Amphitheater in Sunnyvale, California, allow precise manipulation at the fabric corners.

[LEFT] Secondary edge cables provide fabric terminations that lack both the adjustability and the visual clutter of clamped terminations with turnbuckles.

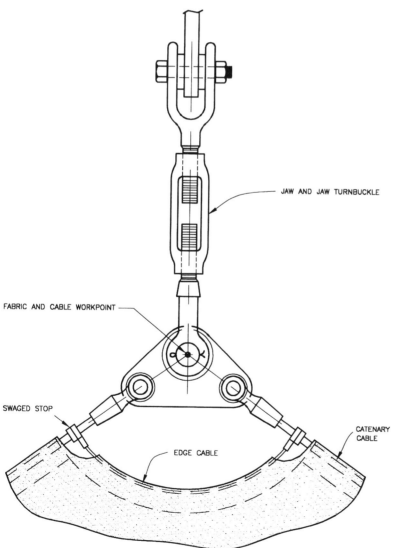

THE TENSIONED FABRIC ROOF

[RIGHT] The broad fabric openings at the catenary cable anchorages on the Parkwest Medical Office Building bring playful interest to the design while allowing the overhead pipe struts to terminate short of the ends of the arches.

[BELOW] Polyester webbing secures the fabric at the mast tops of the Jameirah Beach Hotel's Palm Court canopy. The catenary cables, rather than terminating at the mast, ride in the groove of sheaves secured by shackles.

Source: Paul Roberts; used with permission.

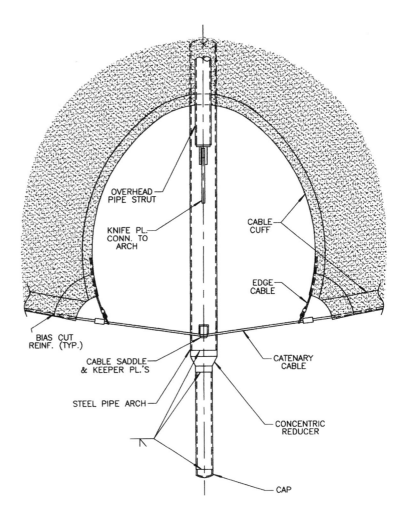

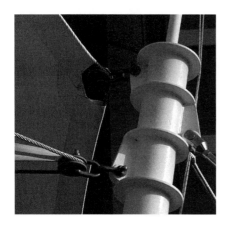

The breadth of secondary cable terminations can be increased to flirtatiously reveal cables or supporting members otherwise hidden from view. In designing the entrance canopy at the Parkwest Medical Office Building in Knoxville, Tennessee, we elected to make the fabric opening created by the secondary cable more than 1 m in depth, enough to expose the connection between the cantilevered arch and the strut that braces it from above (see the drawing above and the photograph on page 37).

A third means of fabric restraint is to connect flat webbing to the fabric alongside each edge cable, sometimes providing a turnbuckle or other mechanism to draw the webbing (and fabric) to the supporting member (see the photograph to the left). The technique has practical application in polyester membranes coated with polyvinyl chloride (PVC), where polyester webbing can be sewn to the fabric. Such webbing generally lacks protective coating, however, and is therefore subject to more rapid degradation from ultraviolet radiation than the membrane itself.

Cable Saddles and Terminations

The primary detailing problems of cables used in tensioned fabric roofs are termination at the cable ends and angular changes at inter-

secting elements along the length of the cable. The different mechanisms that can be used vary in adaptability, economy, and visual elegance.

Eye Terminations

Terminations must transmit cable tensile forces into the supporting structure. They may either be fixed or allow adjustment in cable length, and they must allow the cable to articulate through angle changes about one or both axes. The most economical termination, admirable for its easy field application, is a looped cable eye formed by a thimble and secured by U-bolted clips (see figure above). Such terminations may be adjusted for length prior to installation, and they allow the cable to articulate about both axes. Their application is limited by their inelegant appearance and potential for improper installation, but they are well suited to temporary structures or those with limits on budget or need for sophistication. Their appearance is enhanced when a swaging sleeve is substituted for the cable clips. These sleeves may be shop or field applied on cables of up to 30-mm diameter.

Cable eyes may be interlocked with one another to splice two or more cables together, and the eye ends of one or more cables can be linked with shackles to provide attachment to ear plates welded to the supporting structure. The thimbles used in cable eyes force the cable to a tight bending radius, and the designer must reduce the allowable cable capacity appropriately, generally 10 percent for cables of 25 mm or greater diameter and 20 percent for smaller cables (Crosby 1998).

Swaged and Speltered Terminations

Swaging or speltering of stud end, jaw end, or eye end terminations provides reliable fixed-length cable terminations that are generally both more expensive and more sophisticated in appearance than eyes (see photographs to the right). In swaging, the fitting is clamped tightly onto the end of the cable. Smaller swages can be made in the field, although shop work is preferred.

Spelters are formed by pouring molten metal inside a tapered sleeve as required to fix a cable whose wire ends have been spread open inside the fitting. While speltering may be done in the field, it is generally completed in the shop, where work is faster and more reliable.

Both swages and spelters are available with stud ends (threaded terminations that provide adjustment in length), jaw ends (which attach to a single ear plate with a pin), and closed ends (eyes that

[LEFT] Cable eye ends that employ a thimble and several cable clips provide economical and adaptable terminations for cables with diameters up to 55 mm.

Source: The Crosby Group, Inc.; used with permission

[a]

[b]

[c]

[d]

[ABOVE] Swages with pinned jaws (a) and closed ends (b) are economical for use on cables of 60 mm or smaller diameter, while jaw end (c) and closed end (d) spelters are available in larger sizes. Swages and spelters also are available with male-threaded stud ends (not shown).

Source: The Crosby Group, Inc., used with permission.

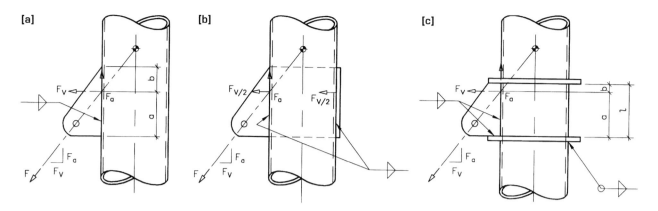

[ABOVE] (a) In designing face-welded ear plates, cable force, F, is resolved into axial (F_a) and lateral (F_v) components resisted through bending of a single mast wall. Sidewall bending is minimized by equalizing the dimensions "a" and "b." (b) Knifing the ear plate through the mast effectively halves sidewall bending. (c) Sidewall bending is effectively eliminated with ring plate designs. Ring plates and the ear-to-ring weld must be designed for the appropriate components of F ($F_v \times a/l$ at the top ring and $F_v \times b/l$ at the bottom ring).

[ABOVE] Turnbuckles may be supplied with either jaw or eye ends.

Source: The Crosby Group, Inc.; used with permission.

[BELOW] There are several varieties of "pipe turnbuckle" that provide cable length adjustment with greater visual elegance and less clutter than conventional turnbuckles.

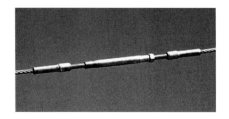

may be secured between pairs of plates or onto clevises). Stud ends may be either fixed or allowed to rotate, depending on how they are attached. Jaw and closed ends allow rotation about a single axis in line with the center of the pin or eye hole, while toggles or pairs of shackles are sometimes added to permit rotation about both axes.

Turnbuckles

In addition to stud ends or adjustable toggles, cable length variation may be provided by splitting the cable into two segments joined by a turnbuckle (see the photograph to the left). Turnbuckles, like cable eyes, are inexpensive, adaptable connectors that typically have a busy and workmanlike visual character that may be poorly adapted to the most elegant structures. Just as speltered or swaged terminations provide a tidy alternative to eyes, conventional turnbuckles are sometimes replaced (at increased cost) by pipe turnbuckles (see the photograph below left).

Designing Cable Terminations

The choice of cable termination must be coordinated with the design of the elements to which the cable attaches. A closed eye may be knifed between and connected to a pair of plates, while a jaw connects to a single plate. The greater simplicity of the latter connection accounts for the much wider use of jaw ends.

The ear plates to which jaw or eye terminations are attached must be sized with thickness and edge radius adequate to prevent both bearing failure at the pin and shear or tension failure on the net section of the plates adjoining the pinhole. Typically, washer-shaped "boss" plates are welded to each side of the ear plate to match its thickness to the width of the jaw and prevent bending of the pin.

For all but the lightest cable forces in combination with stout mast walls, special care must be taken to avoid excessive local bending in the mast wall that can result from ear plate forces normal to the mast axis (see the drawing above). In some designs, it is sufficient to knife the ear all the way through the column so that both walls of the column are engaged in resisting local bending. In more

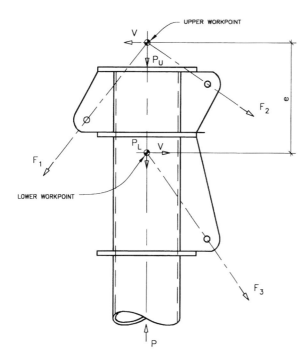

[LEFT] Masts are analyzed by resolving cable forces (F_1, F_2, F_3) into axial and shear components at the mast centerline. The mast is then designed to resist both the total axial force applied by the cables ($P_U + P_L$) and the bending moment associated with the eccentricity between the upper and lower workpoints ($V \times e$).

[ABOVE] Connections such as those at Tampa Airport Hanger No. 1 provide workmanlike and visually unambiguous terminations of both cables and fabric.

heavily stressed connections, however, designers provide circular ring plates around the mast to which the top and bottom edges of the ears are welded. Unnecessary bending moment is avoided, where possible, by laying out the holes for cable pins on the ear plates to bring all cables to a common workpoint at the centerline of the supporting member. There is both economy and a handsome and sturdy honesty to cable terminations that follow these simple principles (see the photograph to the right).

In some structures, the confluence of cables at the top of a mast makes it difficult either to provide the required clearances for cable terminations or to avoid a cluttered appearance. To address these problems in our design of the Weber Point Events Center in Stockton, California (for contractor Sullivan & Brampton), we raised the workpoint for the straight guy and tie-back cables 600 mm above that for the catenary and radial cables that lie in the plane of the fabric (see the drawing above). The double workpoint came at high cost, though, as the resulting bending moment made it necessary to increase the diameter of the larger masts from 324 to 508 mm.

Cable eyes, swages, and spelters each have a different visual character, just as stud, jaw, and closed terminations connote different means of transferring the cable load to the supporting member. Careful designers are cognizant of this as they strive for orderliness in the selection of cable terminations throughout the structure. This order is particularly important where cable terminations are in proximity at mast tops, although different terminations may be successfully combined when their function (e.g., catenaries and tie-backs) is distinct.

Even subtle differences in termination can significantly impact the visual orderliness of a connection design. Catenary cables are often fabricated with a swaged or speltered jaw on one end and a

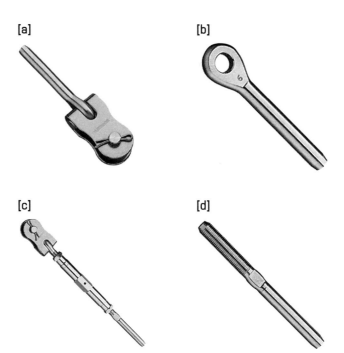

Stainless steel swaged fittings with toggle (a), eye (b), turnbuckle (c), and stud (d) ends provide compact, elegant, corrosion resistant, and more costly alternatives to conventional forged steel fittings.

Source: Ronstan International; used with permission.

swaged or speltered stud end with a clevis jaw at the other, in order to allow the cable to be threaded through the fabric edge cuff. Clevis jaws are larger and of different shape than swaged or speltered jaws of the same capacity, however, and mating the two types of terminations at the top of the same mast may disrupt the visual order of the design. To eliminate this condition, the designer can split the cable into two halves joined with a pipe turnbuckle at the center of the cable length. By threading the stud end of both cable halves and joining them at the turnbuckle, the cable is left with matching swaged or speltered terminations at the two far ends of the cable.

With the exception of the hardware shown above and on the following page, the cable termination hardware shown in the photographs in this section uses forged carbon steel, typically galvanized for corrosion resistance. Such hardware was originally developed for use in industrial rigging applications, where it is subject to impact loading and fatigue due to a high volume of loading cycles. To assure reliability in this demanding load environment, carbon steel swaged and speltered end fittings are generally rated by their manufacturers to provide a high factor of safety (4 to 5) on average ultimate strength.

Terminations are also manufactured using stainless steels, which have greater corrosion resistance than galvanized carbon steels (see the photographs above). Because much of this hardware was originally developed for yachting applications where low weight is important, these fittings use high strength materials that provide compact fittings with a reduced factor of safety of at least 2.2 on working stress loads. This lower safety factor still provides an ultimate strength equal to or greater than that of the cables to which the fittings attach, however, so that no reduction in the ultimate strength of the assembly results. Typically, stainless steel jaw end fittings have

Seven cables terminate in stainless steel jaw ends at a single compact connection at the end of a pipe strut on the Seaworld Whale Pool Canopy.

Source: Claude Centner; used with permission.

narrower throats than those made with carbon steel, such that boss plates are not required at ear plate attachments. While stainless steel fittings are more costly, their reduced mass increases the practicality of complex connections at the same time that it makes all connections more compact and elegant (see the photograph above).

With large cables or geometries that provide very acute angles between cables and masts, ear plates can become high in both their real and visual mass. Refinements sometimes yield connections that are more elegant, more directly expressive of the flow of tensile forces, and that begin to share some of the voluptuousness of the membrane forms themselves. Simple curvature in the edges of ear plates accomplishes this, and further lightness is sometimes achieved by cutting openings in the interior of the plate at areas of low stress. Such paring away of form complicates design and fabrication without offering any compensatory technical advantage, although contemporary plasma-cutting technology minimizes the additional cost associated with such work. Further refinement of cable termination form is achieved when steel castings are substituted for welded plates. Castings gain in economy at heavily loaded connections or where the roof configuration permits substantial reuse of casting forms (Huntington 1993).

Cable Saddles

Special detailing also is required when cables pass without termination over supporting members or intersect other cables, where a saddle is required to guide its changes of angle. The primary considerations in designing saddles are the size of and tensile force in the cable, the cable's range of directional orientation under load, and whether the cable must be restrained from sliding across the saddle

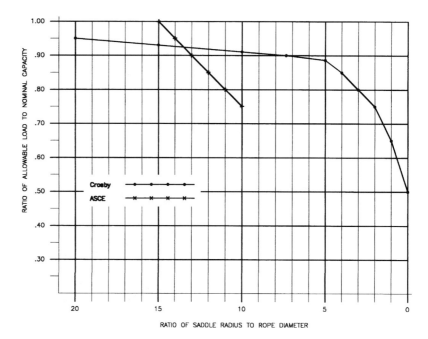

Design criteria provided by the manufacturer Crosby allow for substantially tighter saddle radii than those in the standards of ASCE.

Source: The Crosby Group, Inc., used with permission.

(in order to effectively realize an analytical determined change in cable force in the sections of cable to either side of the saddle).

In general, the large bending radius of structural strand precludes its use with saddles, and the designer will specify more flexible wire rope. The determination of saddle radius is a critical design decision (see the chart above). A radius that is too small will reduce the effective cable strength, while one that is too large yields a saddle of unwieldy size. The American Society of Civil Engineers (ASCE) specifies a minimum saddle radius equal to 15 times cable diameter in order to realize the full cable strength and allows no saddle radius less than 10 times cable diameter, for which it requires a 25 percent reduction in design strength (ASCE 1997).

Conformance to this standard results in saddles of enormous width where the cable passes through a large angle change. Certain industry publications allow much tighter bending radii. The Crosby Group, for example (Crosby 1999), recommends the same 25-percent reduction in cable allowable load where bend radius is only twice cable diameter. I am not aware of problems resulting from the use of such tighter radii, although designers need beware that problems of cable fatigue will occur with small-radius saddles in designs where the cable may work back and forth over the saddles under varying load.

Careful saddle design should consider not only the orientation of the cable to each side of the saddle but also the range of motion through which the cable goes under load, and should ensure that the saddle is configured so that the cable never binds against sharp plate edges. In some cases, the curvature of the cable relative to the orientation of the saddle plate or plates results in the need not only to round the edge of the saddle but also to bend the plate out of plane. Where the cable angle change across the saddle is large, the detail

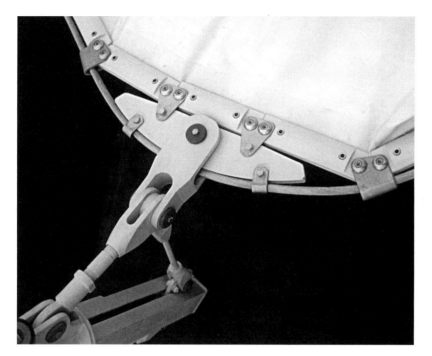

The cable saddles of El Grande Bigo combine elegantly shaped steel and aluminum elements with the judicious use of color. The fabric termination does not provide the adjustability of other connections, such as those of the National Semiconductor Amphitheater shown on page 107.

Source: Canobbio S.p.A.; used with permission.

must provide some form of corner termination for the fabric if a massive saddle is to be avoided (see the photograph above).

In addition to providing an appropriately curved bearing surface for the cable, saddles typically employ some form of "keeper" to prevent the cable from popping out of the saddle during erection or under extreme loading. The keeper may be as simple as a shackle pinned through the saddle plate to capture the cable in position (see the photograph on page 108). The complex requirements of some saddle designs test the skills of good designers to develop connections that not only perform well structurally but also are economical and visually elegant. The best designers configure the structure to avoid complex connection design requirements in the first place.

Where analysis indicates a substantial variation in the cable tension on the two sides of a saddle, some means of restraint must be provided to prevent the cable from sliding across the saddle. Where the force difference is relatively small, sliding resistance may be provided by field bolted saddles designed to clamp the cable between saddle plates. The frictional force provided by clamp plates is difficult to measure and clamped saddles are sometimes difficult to execute in practice. Set screws that bear against the cable provide a simple securement device, but should generally be avoided due to possible local damage to the cable. Sleeves swaged onto the cable to either side of the saddle provide a simple restraint able to resist large differential cable foces.

Cable-to-Cable Connections

At intersections, cables either terminate or are continued over saddles shaped to create the required angle change. In the Hollywood & Highland Ballroom Canopy, the valley cable is terminated at a closed

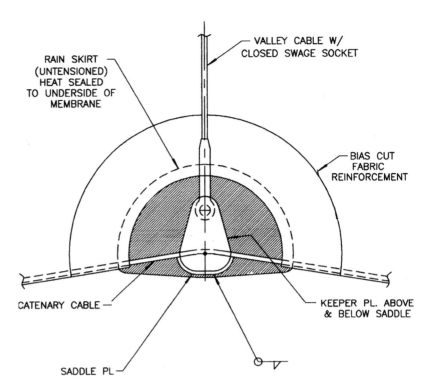

Where cable angle changes are small, as on the Hollywood & Highland Ballroom Canopy catenary, a saddle provides a more compact and economical connection than terminations at the intersection with the valley cable.

swage socket, while the catenary rides across a curved saddle where they meet. The saddle and swage are joined by a pair of keeper plates (see the drawing above). The structural fabric membrane is cut away in the vicinity of the connection; this is required to avoid abrasion of the tensioned fabric on the steel parts. Because project conditions required that water drain over the edge of the structure, however, an untensioned fabric "skirt" was heat sealed to the underside of the fabric to catch and divert any water to the edge of the structure.

By making the catenary continuous over the saddle in the cable-to-cable connection described above, the cost and clutter of cable terminations is avoided. Saddles cannot be used, of course, where the cable must change size at the connection. Large cable angle changes across the connection, particularly where large cables require a large saddle radius, may make saddles unwieldy in size. In such instances, multiple cables may terminate with thimbles and shackles or open swages or spelters onto a shared "flying plate" (see the photograph at the top of the next page).

Mast-Top Connections

Structural Considerations

The mast-peak connection must perform several structural functions. First, the radial tension forces in the fabric must be collected and delivered, without stress concentration, into the mast (see the drawing on page 18). The fabric often necks down to something approach-

[LEFT] This "flying plate" at Stanford University's Avery Aquatic Center gathers three cables into an adjustable tensioning rod.

Source: Don Douglas; used with permission.

ing a point at the top of a mast, and fabric stresses must be transferred over narrow lengths and often at locations confined by cable terminations. Sharply peaked tent cones generally occur only in the mind of the artist, and the designer must satisfy both the structural and visual problems associated with the design of the tension ring.

The mast-peak connection also must provide a mechanism for anchorage of any radial cables that underlie the fabric, through fastening them either to the tension ring or directly to the mast peak itself. Where external cabling is used to stabilize the mast peaks or for some other purpose, it too must be fastened to the mast. Lastly, mast peaks are commonly used as the location for membrane tensioning devices (Huntington 1992).

Fabric termination at the top of the mast may be accomplished at minimal expense by securing a roped edge between pairs of clamp plates that in turn fasten to jaw end turnbuckles attached to ear plates on the mast (see the photograph to the right). The clamp bars are jointed to allow the fabric to articulate around radial or ridge cables coming to the mast peak. These cables are generally terminated either at ears welded to the mast or, where tensioning is done at the mast peak, to a pipe sleeve that rides over the mast.

This approach can provide an inexpensive fabric termination well suited to roofs with highly irregular shapes at the peak, and the turnbuckles provide easy adjustment to accommodate localized irregularities. Care must be taken in their detailing and erection, however, in order to avoid fabric tears caused by stress concentration or sharp edges at the clamp bar joints.

Alternatively, the fabric terminates at a rigid ring constructed with a rolled steel tube or similar member. Plate or tubular steel standoffs may fix the geometry of the ring relative to the top of the

[ABOVE] Clamped fabric edges of the type used at the National Semiconductor Amphitheater provide a readily adjustable and cost-effective mast peak. The photograph shows the temporary rigging used to raise the fabric canopy, and the vertical wrinkle at the right end of the clamped fabric edge indicates a stress concentration that required monitoring during the tensioning process. See also page 107.

The tension ring at Tampa Airport Hanger No. 1 provides termination for both fabric and radial cables, and is tidier and less prone to causing fabric stress concentrations than a simple clamped edge. It is also less adjustable and generally more expensive.

mast, or turnbuckles may allow the ring to "float," thereby permitting field adjustment to perfect fabric or cable geometry (see photograph above).

German designers, led by Frei Otto and his pioneering team of designers, developed a very different approach to the design of the mast-peak connection. By supporting a steel cable loop from the peak of the mast (see photographs on page 41), the top edge of the fabric is attached to the cable in order to distribute its stress over an appropriate length (Huntington 1992). When used in cable net structures, as per the original German application, the cables in the net are fixed to the cable loop with clamps in a manner that allows transfer of the tension in the net into the entire length of the loop. In a pure fabric membrane that terminates in a conventional cable cuff (see the drawing in the middle of page 140), however, the fabric can slip down the sides of the loop in a manner that may cause wrinkling along the sides of the loop and an undesirable concentration of stress transfer at the end of the loop furthest from the mast. Stiff webbing connected to the mast peak and sewn to the fabric alongside the cable may alleviate the problem.

Nonstructural Considerations

The mast peak performs important nonstructural functions as well. Some peaks are designed to allow ventilation of hot rising air through the inside of the tension ring (with or without the addition of a cap

The mast peak baffle used at the University of La Verne mirrors the fabric itself in shape and color.

configured to keep out rainwater), while others are sealed against both air ventilation and water infiltration.

A few structures have employed operable ventilation systems that vary air flow. At the El Monte Resort, our mast top design incorporates an inflatable donut-shaped fabric bladder in the gap between the tension ring and the conical hood. The motor-operated bladder may be filled to seal the gap or deflated to open it for ventilation, as required to suit varying temperature and humidity in the swimming pool enclosure (see page 106).

Mast-peak design must carefully address aesthetics, as well, because the mast peak's positioning at the top of the structure and its function as the point of transference between membrane tension and mast compression makes it a visual singularity whose natural drama ought best complement the inherent bravura of the cone form itself (Huntington 1989).

The nonstructural requirements of the mast peak have been addressed in widely divergent ways on different projects. When installation of the roof was completed by Birdair in 1973, the University of La Verne Campus Center was the first structure to use a polytetrafluoroethylene (PTFE)-coated fiberglass fabric roof. The finished roof gives no visual hint of the transfer of load from the fabric and cabling into the mast because the entire connection is covered by a steel baffle (see the photograph above). The baffle is a cone itself, shaped to continue the curve of the terminated fabric upward

A glazed dome on top of a rigid tension ring, as at Florida Festival, provides a visually distinct skylight in the middle of a sea of fabric.

[TOP OF PAGE] Interior view.

[ABOVE] Exterior view.

from the tension ring to a sharp peak. The mast-peak design reflects a sculptural approach that offers no clue as to the configuration of the structure that underlies it.

While structurally ambiguous, the LaVerne cones are sculpturally appropriate and pleasing. Other approaches have proven less so, as at Florida Festival, where the tension ring and mast peak are hidden by a hemispherical glazed dome reminiscent of the gun turret on a World War II bomber (see the photographs on this page).

Cable loops can maintain the drama of the sharply peaked cone, where desired, while adding the elegant curve of the loop. The designs clearly express the collection of fabric tension and its transfer into the mast. The cable loop has drawbacks when it is necessary to close off the area within the cable loop to provide a weathertight roof, however. On some structures, clear acrylic has been used, but the difficulty of providing a correctly curved edge and maintaining a weathertight seal to the flexible cable must be overcome.

The dramatic impact of the mast peak has sometimes been increased by extending the mast up well beyond the top of the fabric and the tension ring, though the results can appear contrived. At the King Fahd Stadium in Saudi Arabia (see the photographs on page 36), the cylindrical structural masts terminate at the elevation of the tension ring. During construction, immediately following the prestressing of the fabric over the masts, the structure had an elegant and lively beauty. To complete construction, slender mast extensions

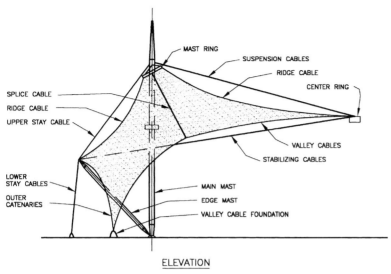
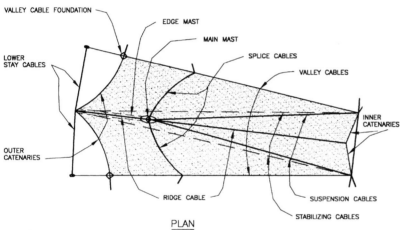

At King Fahd Stadium, the vertical mast and its lengthy extension are misaligned with the resultant of the fabric form, resulting in heavy stay cable loads. Suspension and stabilizing cables equilibrate the flat fabric near the center ring.

were secured atop the tension rings. The dagger-like forms, reminiscent of ancient minarets, are a disappointing addition to the structurally expressive and modern design.

The vertical orientation of the Riyadh Stadium mast and its extension also requires comment. Masts are most efficiently constructed so that they align with the resultant of the accumulated forces at the peak of the fabric. External cabling to the top of the mast serves primarily to stabilize it in the event of a strong wind load or a tear in the fabric. At Riyadh, though, the resultant of the fabric membrane shape pulls markedly inward at its peak while the mast is vertical in order to satisfy nonstructural considerations (see the drawing above); the external guy cables must perform the additional function of resisting the horizontal resultant force at the top of the mast.

A highly successful use of the extended mast was made on the Independence Mall Pavilion erected in Philadelphia for the 1976 United States bicentennial (see the photograph on the next page). The aesthetic has a strong structural element: the exposed cabling carries its tension from the fabric to the mast as palpably as a halyard raising a mainsail (Huntington 1992).

The lean tracery of the mast peaks of the Independence Mall Pavilion provides a daring complement to the form of the fabric canopy itself.

Source: Geiger Engineers; used with permission.

Mast-Base Connections

The connection at the base of the mast requires only the termination of a single compression member without the complexities of fabric and cable termination that occur at the mast peak. While the technical requirements of the mast base are simple, it too is an important visual element in the design and one whose refinement is made more important by its proximity to viewers at grade level.

Besides force level, the fundamental design variable in the mast-base design is the "degree of freedom" of the connection. Depending on its configuration, the mast may be fixed in a manner that prevents rotation about its base, or it may be allowed to rotate about one or both axes of the mast.

Fixed bases have the advantage of allowing the mast to stand vertical during erection, like a flagpole, without benefit of any guying cables; and they are appropriate where lateral movements at the top of the mast do not induce excessive bending in the mast or where it is not possible to use guys or other devices to resist lateral loads at the top of the mast. The most prominent of all fabric structure fixed-mast bases is that on the 170-m-high tower from which the Montreal Olympic Stadium membrane is suspended (see the photograph at the top of the next page). The massiveness of this tower relates less to its structural function than to its architectural roles of enclosing space and giving form. It is nonetheless a heavy-handed sculptural element, ill-suited to the visually delicate mem-

122 THE TENSIONED FABRIC ROOF

brane that it supports. On masts of modest size, fixed bases are typically achieved by welding a steel mast directly to its base plate and securing the base with broadly spaced anchor bolts to a foundation designed against overturning.

Single-degree-of-freedom bases are useful where the post is expected to rotate primarily about a single axis, as at the perimeter of a structure where the roof is tensioned by shortening tie-back cables in a manner that displaces the top of the post outward (see the drawing at the top of page 124). In a manner analogous to a cable jaw end, a single-degree-of-freedom mast base can be achieved by knifing a vertical plate welded to either the mast or the base plate between a pair of plates welded to the opposing element (see the drawing at the bottom of page 124).

Mast bases with two degrees of freedom allow the top of the member to displace freely about both its axes, as is required when such displacements may occur either during erection or in service. Bases that are true "pins" may be created with spherical bearing plates. Mast bases that are stable over wide angle changes about both axes are sometimes required for deployable roofs, and they may be created with mechanical gimbals.

Adequate rotational capacity can generally be obtained with a reduction in both expense and visual expressiveness by placing a pad of elastomeric or other flexible material between the mast bottom plate and the base plate resting atop the foundation. In our design for

Structural function is nearly incidental to the design of the sculptural tower with massive fixed base that supports the delicate membrane of the Montreal Olympic Stadium.

Source: Jörg Schlaich; used with permission.

THE TENSIONED FABRIC ROOF 123

[RIGHT] On the Palm Court Canopy of Dubai's Jameirah Beach Hotel, we relied on an articulating mast base and tie-back cable turnbuckle to introduce tension to the canopy. The cover pipe, provided for visual purposes only, is slid into position to cover the articulating mast base after the mast is erected, but a gap is provided at its bottom to allow for deflection in service.

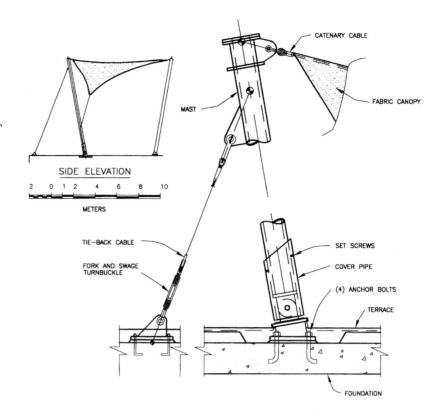

[RIGHT] A simple and visually direct single-degree-of-freedom mast base was created at Hollywood & Highland using three "knife plates" linked with a pin to allow rotation.

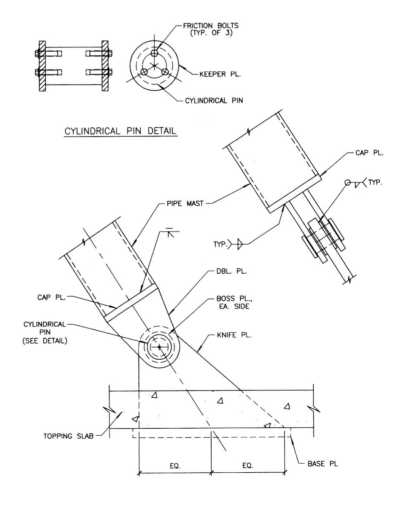

124   THE TENSIONED FABRIC ROOF

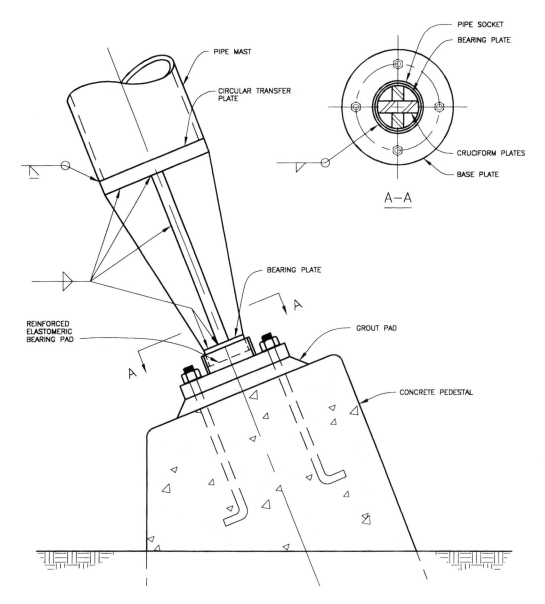

the Weber Point Events Center, we gave visual expression to the pinned mast base by tapering the cruciform mast base to a small diameter at its bearing on the base plate (see the drawing above) If such mast bases rest on the base plate without positive securement to it, it is important to ensure that the mast has compressive load throughout erection and under all load cases, so that the mast cannot pop out of its bearing.

Masts may be tapered to a circular bearing plate in a socket formed from a short section of pipe. The bearing plate is made as small as possible without overloading the flexible bearing pad.

Pretensioning Mechanisms

Pretensioning mechanisms range from simple clamped fabric edges to adjustable cable anchorages and telescoping masts. Their design must provide a measurable range of motion that results in the fabric being tensioned a predictable amount, and in a manner that both provides a smooth membrane surface and prevents excessive flutter of the fabric under wind loading. Well-designed details result in tension structures that are straightforward to erect and, when left

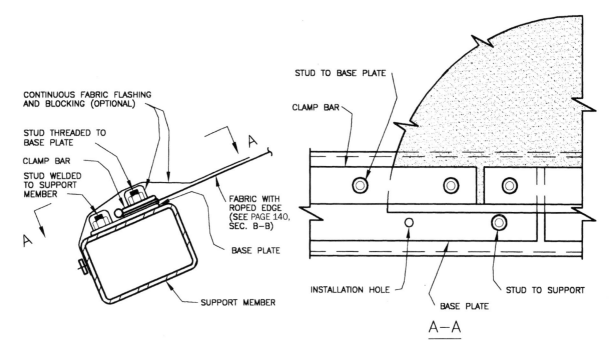

Double plate clamped edge.

exposed in the finished roof, are elegant and expressive of the flow of forces in the tensioned roof.

Direct Tensioning of Fabric

The simplest tensioning mechanisms employ simple, clamped edges similar to those used in some awnings. The fabric may have a roped edge that is secured between metal plates, as discussed earlier in this chapter. Single aluminum clamp bars (see the drawing on page 105) do not readily adapt to tensioning. To facilitate tensioning, other designs provide a base plate beneath the fabric that has holes for securing come-alongs or other mechanisms that pull the fabric out to its final position (see the drawing above).

Edge-tensioning systems are useful on small structures with low prestress forces, and on those with no masts or perimeter catenary cables that may be adjusted to provide tensioning of the fabric. In comparison to other tensioning mechanisms, they have the advantage of simplicity, but they require that tension be applied manually to the fabric at close intervals and generally provide no allowance for adjustment.

A refinement of the basic clamped edge system that does allow adjustment was developed for the entrance canopy to the Marin Technology Center in San Rafael, California. The perimeter clamp bars are secured to turnbuckle jaw ends that are adjusted with a nut to provide appropriate fabric tension (see the drawing on the next page) (Huntington 1997).

Cable-Tensioning Mechanisms

While the above systems rely on direct application of tension to the continuous edge of the fabric, greater simplicity is generally achieved

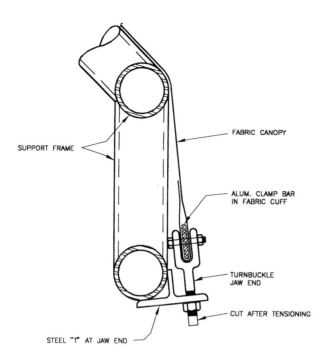

Adjustable edge clamping.

by providing movement at one or a few points on the supporting structure. Often this is done by providing adjustment in cables. In one system, shortening perimeter tie-back cables by adjustment of a turnbuckle pulls back on perimeter support masts; this in turn tensions the catenary cables at the edge of the fabric and draws the fabric itself into tension (see the upper drawings on page 124).

Tie-back cable-tensioning devices are readily applicable in structures where the fabric terminates in a perimeter mast system, and where the obstruction at grade created by the tie-back cables is acceptable. In other applications, tension is applied directly to the catenary cables. Because the entire roof may be tensioned by making adjustment at only a few locations, both types of cable tensioning systems have an advantage in ease of erection over direct tensioning of the fabric. Cable tensioning may lead to economies in the supporting structure as well. Direct fabric-tensioning devices generally anchor to perimeter beams that carry load inefficiently in bending, while cables generally anchor at the intersections of supporting members so that the supporting members are loaded in direct tension or compression. Furthermore, by freeing the edge of the fabric roof from the constraints of the rigid perimeter members that are required to facilitate clamping, the catenary edges of these roofs begin to provide the clear elegance of curving lightweight construction that distinguishes fabric tension structures from awnings (Huntington 1997).

Where a continuous edge is dictated by geometric considerations, such as the need to provide a straight line interface at the juncture of the membrane with grade or a wall, designers sometimes maintain the erection advantage of catenary edges by infilling the parabolic gaps between catenary and rigid edge with fabric closure panels. These "nonstructural" membrane elements are lightly

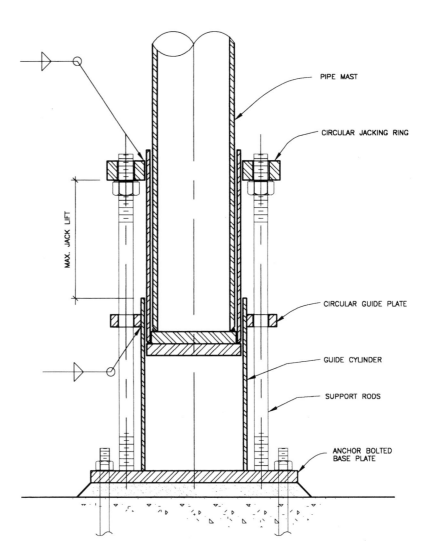

Our design for Buchser School provides jacking at the mast base. In designs of this type, the engineer must assure that the support rods are stout enough to avoid buckling under compression.

stretched out to the continuous edge following tensioning of the primary membrane and are protected from heavy stress under load by the stiffness of the cable.

Mast-Tensioning Mechanisms

The structural engineer might characterize cable-tensioning systems as mechanisms that apply tension to the fabric by reducing the length of tension members (cables). Mast-jacking systems, conversely, apply tension to the fabric by increasing the length of compression members (masts or struts). Like cable tensioning, mast jacking provides a means for tensioning a large area of fabric by making adjustments at a few discrete locations, and it shares applicability to structures with catenary edge cables. Like direct fabric-tensioning systems, though, mast jacking also is useful on structures with rigid perimeter elements.

Jacking mechanisms on masts are typically placed at either the top or bottom of a vertical interior mast. When placed at the bot-

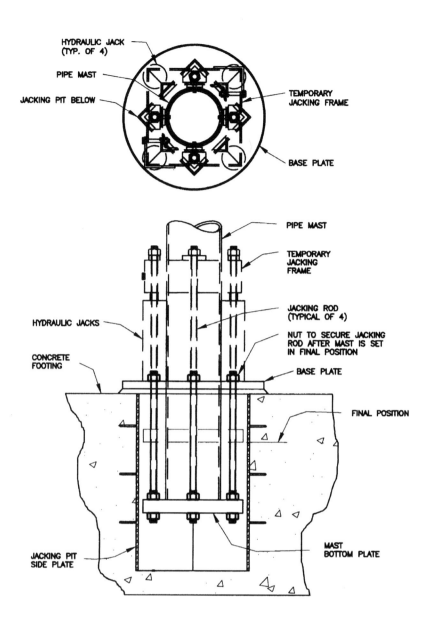

The mast bases of the Chang Sha Amphitheater are recessed into the foundation to provide an uncluttered condition at grade.

tom, they can give a machine-like expression of the means by which the structure is erected or tensioned, with adjusting bolts and pins or bearing mechanisms often left exposed as a muscular reminder of the large forces exerted to raise the mast and pretension the fabric. In my structural design for a shade canopy at Buchser School in Santa Clara, California, for example, the mast sits inside a cylindrical pot that is raised into position with portable hydraulic jacks, then secured with four large threaded rods (see the drawing at the top of page 128). In detailing such a mast base, the designer must assure that the support rods are sufficiently stout to avoid buckling when the mast is raised to its final position and in full compression.

Bottom-of-mast mechanisms may result in a bulky element at grade that obstructs use of the building space and detracts from the structure's visual lightness. In our design for an amphitheater in Chang Sha, China, we overcame this problem by founding the base of the mast inside a pit in the mast footing (see the drawing above).

THE TENSIONED FABRIC ROOF 129

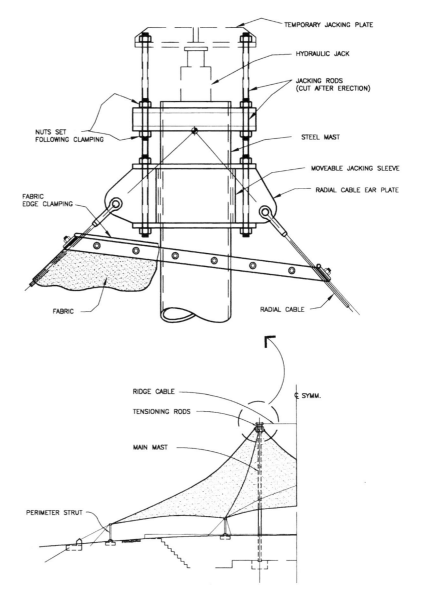

[ABOVE] Mast-base jacking systems like those provided at Tampa Airport Hanger No. 1 are workmanlike and expressive of both the tensioning process and the flow of tensile and compressive forces between structural elements.

[ABOVE RIGHT] Mast-top jacking systems, as employed on the National Semiconductor Amphitheater, can simplify the transfer of loads and eliminate the need for multiple jacks. However, they necessitate that the jacking be performed by workers supported high above grade. The same connection is shown at the beginning of prestressing in the smaller photograph on page 117, with temporary cables in place to winch the fabric up to near its final position before the hydraulic jack and jacking rods are used for final tensioning.

Both the mast-jacking frame and the hydraulic jacks used to raise the base of the mast are temporary elements that are removed following pretensioning, so that the base of the mast is left unobstructed at the completion of construction. When skillfully designed, though, exposed bottom-of-mast jacking mechanisms add a dynamic visual element to the design, with the exposed operable structural members expressing the upward thrust of the mast and the application of tension to the roof itself (see the photograph above left).

Moving the tensioning system to the top of the mast provides another escape from bulky ground-level mechanisms, although the erection procedure is made more complex by forcing workers operating the jacking mechanisms into crane baskets high above grade. Whereas bottom-of-mast systems may require multiple jacks, only a single hydraulic jack is required atop the mast cap plate in order to raise all of the cables terminating at the sleeve surrounding the mast (see the drawings above). A temporary jacking frame or plate is again employed to provide a bearing for the jack.

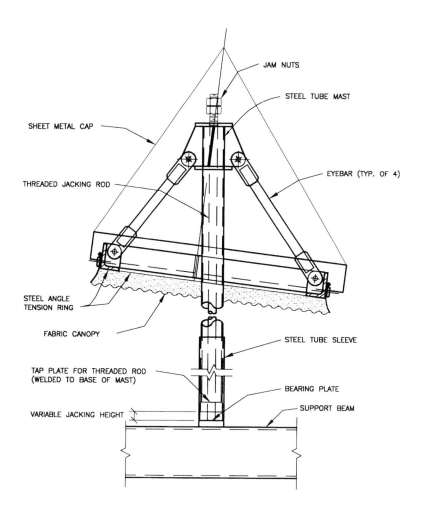

The small scale of the entrance canopy for National Semiconductor made it possible for us to design the mast-jacking mechanism using a single internal rod over the height of the mast. The uncluttered detail is in keeping with the geometric simplicity of the canopy.

In smaller structures, the principles of mast jacking can be used in simple mechanisms that tension the cables and fabric by the turn of a bolt or threaded rod without benefit of jacks. In engineering an entrance canopy for National Semiconductor (see the drawing above and the photograph on page 26), we used a threaded rod that extends inside the mast for its full height and is screwed tight to elevate the mast-top tension ring at which the tensioned fabric terminates. Horizontal or sloping struts can be just as effective as vertical masts in providing the source for a tensioning mechanism. A simple threaded adjusting device elongates the catenary cables to produce tension in the fabric of our umbrella canopy at Jameirah Beach Hotel (see the drawings on the next page).

Tensioned fabric structures derive their distinctive appearance, stability under load, and even their name from the presence of consistent pretensioning in the fabric membrane. The mechanisms used to apply this pretensioning are, by their nature, key design elements. With the selection of the appropriate type of mechanism and cleverness in its execution, the roof designer is well down the road to realizing the goals of easy erection, reliability, and dynamic beauty that are emblematic of the best tensioned fabric structures (Huntington 1997).

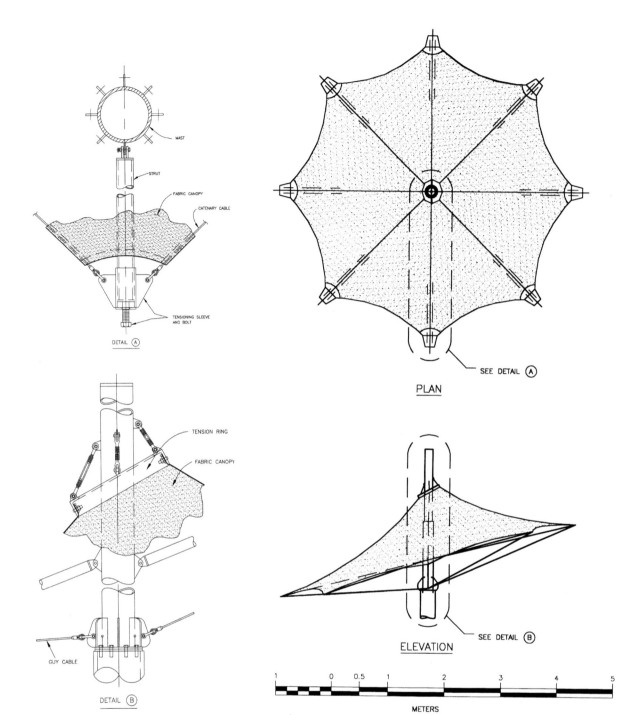

On a small roof canopy at Jameirah Beach Hotel, the fabric membrane and cables are tensioned by the adjustment of a bolt at the end of each strut. The detail is elegant and effective.

7 Fabrication and Erection

The preceding chapters outline the relationship of materials, structural design, and detailing to the performance, durability, and appearance of tensioned fabric roofs. No roof can be successful without expert fabrication and erection, however, and designers must anticipate the methods and constraints of the construction process and incorporate this consideration into each step of their work. The criticality of this is illustrated by noting that no other construction type calls upon field crews to remove a single fabricated piece from a crate, lift it high into the air, and stretch it without damage or mishap into a final and precise position covering 1,000 m^2 or more.

In discussing the process of designing and constructing a tensioned fabric roof, the normally clear boundaries between the work of designers and builders quickly blur together, for reasons touched upon above and described in greater detail in Chapter 9. Items such as seam location and patterning are critical to the appearance and performance of the roof, and the designer must consider them early in the design process. They are most often the responsibility of the roof contractor, however, and are therefore included in this chapter on fabrication and erection.

In the sections that follow, each step of the fabrication and erection process is considered in turn, beginning with the important considerations of the architect and structural engineer during the design process.

Fabric and Seam Selection

Engineers input the stiffness properties of their preliminary fabric selection to the membrane analysis in order to verify that calculated stresses do not exceed the allowables for the proposed material.

As seen at Sherway Gardens Mall in Etobicok, Manitoba, the pattern of opaque seam bands in a field of translucent fabric is an important visual element from inside the structure during the day [ABOVE] and from outside at night [BELOW].

Source: Birdair, Inc.; used with permission.

Then, before proceeding to the generation of a pattern for the fabricated roof, they determine the fabric's "compensation" properties: the amount that the pieces of fabric must be reduced in size when they are cut so that, under prestress, they elongate to the proper size.

Experience demonstrates that compensation test values vary substantially between rolls of nominally alike fabric produced by the same manufacturer. In critical installations, such as buildings with sharply curved surfaces that employ stiff fiberglass fabrics, the engineer will consider the tested compensation value of individual batches of fabric in creating his pattern. In others, he may rely on "typical" or average compensation values for a given fabric product.

Thorough engineers also consider the width of the rolls of fabric to be supplied for the roof. Polyesters coated with polyvinyl chloride (PVC) are typically limited to widths of 1.5 to 2.0 m, a dimension that results in fairly closely spaced seams and the opportunity to shape the membrane with a high level of precision.

[LEFT] The characteristic upward sweep of seams in the cone roof of the Park City Mall in Lancaster, Pennsylvania, draws the eye up with it.

Source: Birdair, Inc.; used with permission.

[BELOW] Continuous seams emphasize the longitudinal axis of a repetitive arch roof such as at Buena Ventura Shopping Center in Ventura, California.

Polytetrafluoroethylene (PTFE)-coated fiberglass fabrics are available in widths up to 4 m, too wide to articulate tightly curved shapes but appropriate for a structure like the large and softly curved Hajj Terminal, where the wide seam spacing helped to reduce fabrication costs and limit opportunities for seam failure. Where curvatures dictate closer seam spacing, the fabricator will employ narrower rolled goods. In any case, the roof fabricator, like a skilled tailor, always seeks to space seams and lay out individual patterns on the fabric roll in a manner that minimizes waste.

The orientation of seams is usually determined during the design process, based on several considerations. The lapped fabric at seams causes a reduction in the membrane's translucency that, when viewed from the inside under daylight conditions, results in clearly visible dark seam bands. A similar effect can be seen from the outside when artificial lights inside the building shine through the fabric at night (see photographs on page 134). Just as a fashion designer considers the tendency of vertically or horizontally oriented stripes on clothing fabric to make the wearer appear tall and slender or short and stout, the fabric roof designer must consider the way in which the repeated seam bands may draw the eye up to the confluence of seams at the peak of a cone or exaggerate the length of a long canted arch atrium cover (see photographs on this page).

In better structures, engineers also consider the visual effect of seams where fabric assemblies join to each other. Ideally, the seams in each assembly align with those on the opposing side of the assembly joint (see the photograph at the top of the next page) to form a chevron pattern. Without carefully coordinating the width of individual fabric strips and the angle at which the seams meet the assembly edge on each side of the joint, however, the seams to either side will misalign

[ABOVE RIGHT] The patterning and alignment of fabric seams fits hand in glove with the religious imagery of the Good Shepherd Church in Fresno, California.

[RIGHT] Asymmetry across the panel assembly joint may make alignment of the seams difficult without high fabric waste.

Source: Robert Reck; used with permission.

(see photograph above). Unfortunately, alignment of all seams is sometimes inconsistent with the goal of minimizing material waste.

The choice of seam direction also determines the orientation of warp and fill in the fabric. Warp fibers are laid parallel to the seams and generally elongate little relative to the fill during prestressing. Because of this, seam orientation must be coordinated with the proposed erection procedure and the sequence in which perimeter fastenings are made. This is covered in greater detail in the section on erection.

While seams are generally intended to develop the full strength of the base fabric, the fabricator also may try to orient seams so as to minimize the stress in the fabric perpendicular to the seam and to minimize the hazard of seam failure. This consideration may be critical with any glued seam that is subject to "creep" under load.

Fabricators also observe limits on the width of fabric between seams in sharply curved structures, as noted above. On broadly

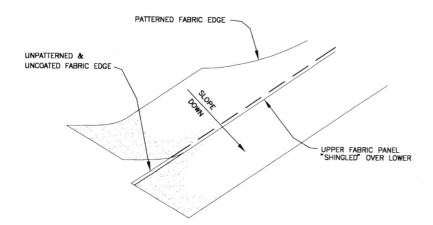

To simplify fabrication, alternate panel edges terminate at straight edges with no topcoat. At these locations, neither cutting of the fabric nor topcoat removal is required.

curved forms, conversely, the fabricator may join two or more strips of fabric with a straight seam, cutting only the perimeter of the doubled fabric panel to a pattern in order to reduce fabrication costs. This approach has a second advantage when employing PVC-coated polyesters or other fabrics with unweldable topcoats. The topcoats generally must be removed at seam locations in order to provide proper adhesion. Fabric manufacturers typically leave a seam width of fabric uncoated at one edge of the roll, however, so that patterning only alternate seams reduces the number of locations requiring topcoat removal.

In addition to deciding seam orientation and determining the locations of straight and patterned seams, the fabricator must also decide which piece of fabric is to lay over the other at each seam. In a conical form with radial seams, the choice is of no consequence; but in a saddle shape where seams may traverse the slope of the finished roof, seams are best oriented with the higher strip of fabric laid on top of the lower, so that the seam sheds water (and dirt) in the manner of a shingled roof (see drawing above).

Just as important as determination of the heat-sealed or other shop-seam locations are decisions about field-seam locations, as required to accommodate the proposed erection procedure and the practical limits on size of fabric panel that can be shipped or safely erected. Small roofs are generally shop fabricated in a single piece, while larger ones are typically field joined at selected cable locations.

Air-supported roof membranes (and some tension structures) are wound onto rolls for shipment, then rolled into place over supporting members or cables in the field. The fabric can be rolled in lengths of 100 m or more, but the width of shop-fabricated panels in these

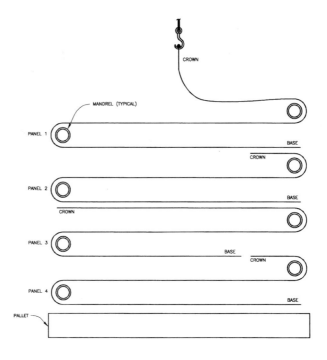

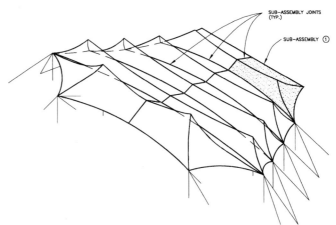

[ABOVE RIGHT] Fabric is laid on pallets and rolled over mandrels, with individual fabric panels configured so that they can be readily lifted by the crown and placed in the required order.

[RIGHT] The computer model used for shaping, analysis, and patterning of the Chang Sha Amphitheater mirrors the form of the finished structure. Sub-Assembly 1 is the focus of the drawings on pages 139 and 140.

structures is limited by the practical length of roll that can be shipped, typically about 15 m.

On most tension structures, though, the fabric is rolled gently over mandrels and laid into crates or over pallets (see the drawing at the top of this page), making wider panels possible. Limits on both the economical lifting capacity of equipment and the size of unprestressed fabric that is safely exposed to winds during erection usually limit shop fabrications to 2,500 to 3,000 m^2 area.

The amphitheater that my firm engineered for the city of Chang Sha, China, illustrates some of the basic principles of seaming design. The roof is a ridge and valley system with an area of approximately 4,500 m^2 (see the drawing above and the photograph on page 34), constructed with a white PVC-coated polyester with polyvinyl fluoride (PVF) topcoat. The erectors had a large crew of workers but only minimal equipment to lift or pull the fabric, so the membrane was divided into ten shop-fabricated sections (see the drawing at the top of page 139) in order to facilitate erection. Each of the ten "sub-

138   THE TENSIONED FABRIC ROOF

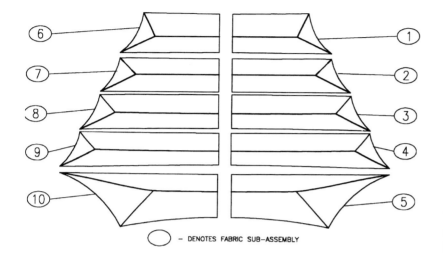

The Chang Sha Amphitheater's ridge and valley roof was divided into ten shop-fabricated sub-assemblies.

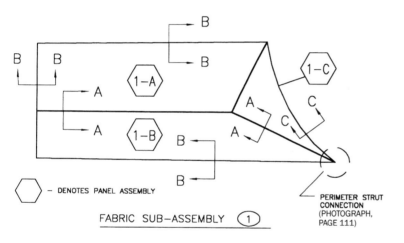

[LEFT] The "sub-assembly" is in turn assembled from three "panel assemblies" that are joined in the shop with welded seams (Section A-A, the middle drawing on page 140). Each of the shop-fabricated sub-assemblies has a field-joined clamped edge (see the drawing on page 102) or catenary cable edge (Section C-C, the bottom drawing on page 140).

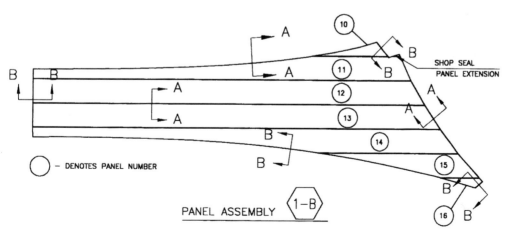

[ABOVE] Panel assemblies are constructed of fabric strips oriented to minimize waste and heat sealed in the shop. Sections A-A, B-B, and C-C are shown in the drawing on page 140.

THE TENSIONED FABRIC ROOF   139

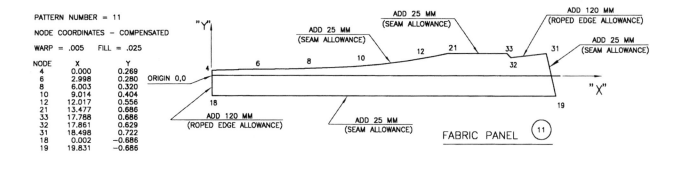

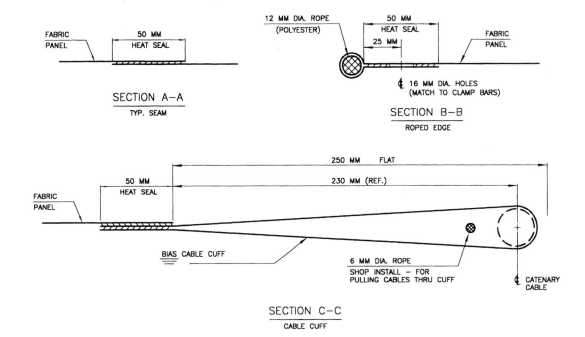

[TOP] The geometry given to the fabricator was "compensated" by reducing it in size 0.5% in the warp and 2.5% in the fill in consideration of its elongation under prestress. The geometry correlates to fabric worklines at the center of catenary cables or roped edge joints. Seam or roped edge allowances must be added to the dimensions shown in order to find the actual cutline for the fabric.

[MIDDLE] Joints between individual fabric panels are seamed as in Section A-A, while panel assemblies terminate at either roped edges (at edges of the structure joining rigid supporting elements or where assemblies are joined to each other) or cable cuffs (where the roof terminates at a catenary cable).

assemblies" (see the drawing in the middle of page 139) is, in turn, divided into "panel assemblies" (see the drawing at the bottom of page 139) composed of several widths of shop-welded fabric.

Because the membrane has only modest curvature between the ridge and valley cables, the fabric required patterning at only alternate seams between fabric strips (see the drawing at the top of this page). The unpatterned edge (from Nodes 18 to 19, at the bottom of the figure) takes advantage of the fabric strip's single un-topcoated edge.

The three typical fabric termination details used at Chang Sha (see the drawing above) reflect the fabrication and erection strategy. Heat welds (Section A-A) are used at all shop seams, including those over the ridge cable, while roped edges (Section B-B) are provided for the field joints at each valley cable. (The complete field assembly of the latter is shown in the drawing on page 102.) At the curved free edge of the fabric canopy, a cable cuff (Section C-C) provides a pocket for the edge catenary cable.

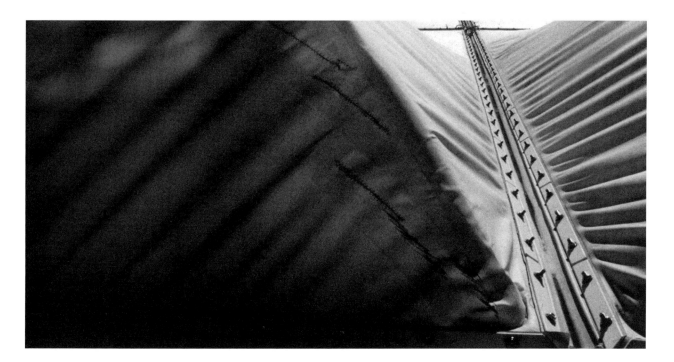

Even a small and simply shaped membrane can suffer disastrous wrinkling and tearing as a result of poor patterning.

Patterning

Having set out the properties of the selected fabric and determined the configuration of seams, the designer must next create the patterns and details by which inert rolls of woven goods are transformed into fabric architecture. The generation of these precise plans represents a marriage of old crafts with leading-edge computing skills — the fabric-working techniques of America's Cup sailmakers joined to those of a traditional tailor.

Until the early 1970s, fabric patterning, like the membrane analysis work discussed in Chapter 5, was done largely using hand computations and empirical wisdom. Under the earlier analysis regime, structurally reliable roofs were generally ensured by making iterative steps forward in both shape and scale. Newer patterning techniques, like advanced analytical methods, have made reliable behavior in innovative roofs a reality. Recalling the Birdair "no patches, no wrinkles" policy, it might be said that, if newer analytical techniques are responsible for the accurate stress analysis that prevented tears (and patches), it is contemporary patterning that has yielded the precise shapes that prevent wrinkling (see the photograph above).

The mathematical model used in the patterning is a mesh of elements that approximate the shape of the finished roof. If the model used for the original shape-finding and analysis is sufficiently detailed and accurate, and if the edges of the elements in this model have been laid out along seam lines, the analysis model may be used for patterning, as well.

Typically, however, the pattern model will include geometric and other refinements to the analysis model. Where applicable, these

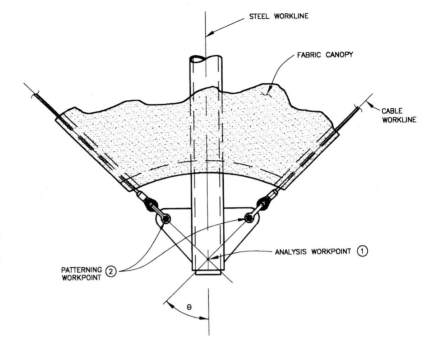

[RIGHT] For analysis, providing a common workpoint (1) at the intersection of cable worklines generally provides sufficient accuracy. For patterning, however, workpoints (2) must be defined at the pin of each cable termination in order to obtain accurate cable lengths. In the pattern model, the cable terminates at a pin-ended link between workpoints 1 and 2 so that the angle 2 between steel and cable worklines is output by the pattern.

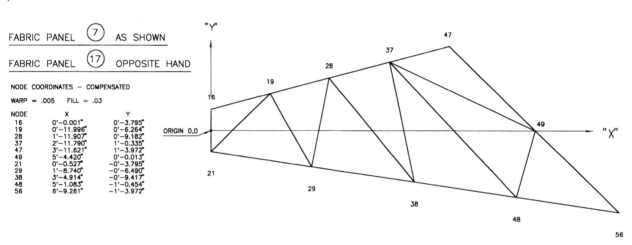

FABRIC PANEL ⑦ AS SHOWN
FABRIC PANEL ⑰ OPPOSITE HAND

NODE COORDINATES – COMPENSATED
WARP = .005   FILL = .03

NODE	X	Y
16	0'–0.001"	0'–3.795"
19	0'–11.996"	0'–6.264"
28	1'–11.907"	0'–9.182"
37	2'–11.790"	1'–0.335"
47	3'–11.621"	1'–3.972"
49	5'–4.420"	0'–0.013"
21	0'–0.527"	–0'–3.795"
29	1'–8.740"	–0'–6.490"
38	3'–4.914"	–0'–9.417"
48	5'–1.083"	–1'–0.454"
56	6'–9.261"	–1'–3.972"

[RIGHT] The curving form of the membrane is modeled by a network of triangular elements which, separated along seam lines and laid out flat, define the geometry of an individual pattern.

will include consideration of the pin location of individual cables, rather than the common workpoint of multiple members (see the drawing at the top of this page).

From the model's surface grid, strips are isolated whose edges correspond to seams and which are composed of triangular elements much like the flat facets between folds in a long strip of paper (see the drawing in the middle of this page). Pressing the facets down flat into a single plane, the edges of the resultant two-dimensional shape define the edges of a fabric element and correspond to the seam lines in the finished roof. After all elements have been cut, they are rejoined along their edges and, in the process, take on the form the roof will have when prestressed into its finished shape (Shaeffer 1996).

Using a sufficiently fine mesh, the engineer can approximate the shape accurately enough that, after compensating for the elongation under prestress, the fabric pattern may be transposed directly to fabric cuts. The importance of accurate patterning and compensation is

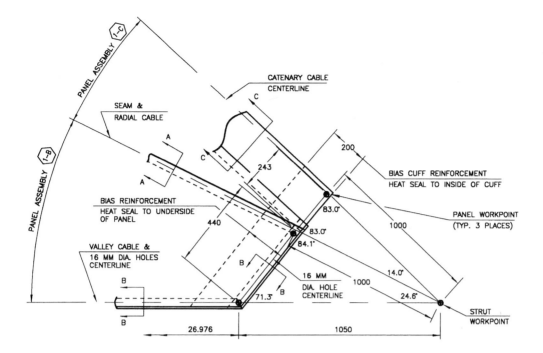

demonstrated by a simple example. If fill compensation is determined to be 2 percent, the fabric will be cut to 98 percent of its finished size in the fill dimension. If an error in patterning results in the fabric being 2 percent *oversized*, the fabric will have no prestress when clamped at its perimeter. If, on the other hand, it is 2 percent *undersized*, prestress may be double the predicted amount, causing difficulties in erection and potential overstresses or tears.

Extra care must be taken in patterning at the edges and, particularly, the corners of fabric that are formed by cables, arches, mast peaks, or other discontinuities. These areas tend to be not only highly visible but also likely locations of highly articulated curvature, stress concentration, abrasion, or wrinkling. Care must be taken to provide accurate patterning at these locations, and supporting elements must be detailed to protect the fabric from sharp corners or edges that might lead to tears. Engineers typically prepare detailed drawings of corner conditions, showing such elements as the faceted shape of the fabric edge to conform to edge clamp pieces and webbing or doubled fabric thicknesses to counter stress concentrations (see the drawing above).

Cutting and Seaming

Fabricators have traditionally converted the engineer's pattern into cut fabric by laying out the fabric on the shop floor, marking out the pattern's node coordinates on the fabric by hand, and connecting them together into the curved fabric cutlines. More recent technologies have allowed fabricators to log an input file of the engineer's pattern into the computer that controls an automated cutting machine. While cutting of the overall fabric panel shape may be automated,

The fabric is carefully detailed and reinforced at the connection to a perimeter strut. The detail incorporates each of the fabric terminations shown in the drawing in the middle of page 140: a seam (Section A-A) at the juncture between panel Assemblies 1-B and 1-C, a roped edge (Section B-B) at the juncture with Sub-assembly 2, and a cable cuff (Section C-C) at the catenary cable. A second roped edge is used on the right-hand edge nearest the strut workpoint, so that turnbuckles attached to both the fabric edge clamps and the strut can secure the membrane against creeping away from the strut.

[RIGHT] Workers align PVC-coated polyester fabric strips under a metal bar sized to create the required width of radio-frequency–welded seam.

[BELOW RIGHT] Full sealing at changes in the layering of fabric is maintained by the use of specially shaped sealing bars.

[BELOW LEFT] Cable cuffs and roped edges are elements of the special detailing required to complete fabrication.

Source: These three photographs were taken at the shop of Sullivan & Brampton; San Leandro, California.

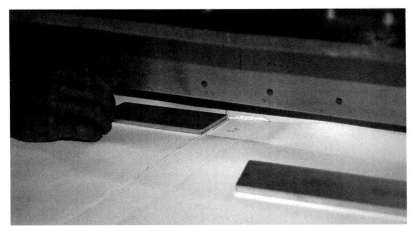

the detail drawings at fabric discontinuities, discussed above, are generally associated with substantial hand detailing in the shop.

Fabric seams can be made in the shop using a variety of technologies. Traditional sewn seams are still used on some structures, although not with fiberglass-based fabrics because of potential damage to the brittle yarns from the needle. The holes created by the needle also create potential openings for water leakage, and sewn seams should therefore be sealed or covered with a strip of fabric where watertightness is required.

Cemented seams also are used in certain applications, but caution is required in their use. A number of early silicone-coated fiberglass structures suffered discoloration at seam lines and, worse, slow creep under load that eventually led to peeling apart of the seams.

The majority of shop seams for architectural fabric structures built today are made using some form of heat sealing, a process that uses a combination of heat and pressure to fuse two pieces of fabric together. In some cases, the coatings of the fabrics are melted together. With PTFE-coated fiberglass, a strip of fluorinated ethylene propylene (FEP) is sandwiched between the two layers of fabric. The FEP, a Teflon® formulation with lower melting temperature than PTFE, melts to the coatings of the two fabric layers under heat to form the bond. While various types of heat sealers are marketed, the most common use radio-frequency energy and dielectric heating to fuse the fabrics (see the photograph at the top of page 144). Because the heat-sealing bar must fit snug against the fabric to make an effective seal, the fabricator uses special sealing bars shaped to fit the changes in fabric assembly thickness (shown middle-right of page 144). Following the seaming of individual fabric panels into fabric sub-assemblies, the fabricator installs reinforcement straps, cable cuffs, edge ropes, and other detailing (shown middle-left of page 144).

The degree of care taken in handling of the fabric on the shop floor is critical to preventing soiling, weakening of the material due to sharp folds, and tears. The shop floor must be kept scrupulously clean during the work and the fabric clear of contact with fluids, grease, or chemicals. Workers may need to step on the fabric to complete certain work but should wear soft-soled shoes when doing so and avoid stepping on folds in the fabric. Special care must be taken in packing the fabric for shipment, as well. The risk of damage to the fabric in the shop is partially a function of the material choice, with PTFE-coated fiberglass generally both most vulnerable to damage caused by folding and most resistant to soiling.

Fabrication of Ancillary Elements

Construction of a tensioned fabric roof requires the fabrication of numerous elements in addition to the membrane. Supporting structural members, typically steel, are usually heavy and large relative to the packed membrane. Since the requirements of their manufacture are similar to those of some other steel structures, they are typically fabricated and erected by a subcontractor proximate to the building site (see the photograph on the next page). Gaskets, clamps, and similar accessories are both more compact and more specialized and are therefore typically fabricated or procured by the fabric roof contractor itself.

The fabrication of cables must be closely coordinated with the patterning process, which not only provides the precise shape of fabric cuts but also determines the exact fabrication lengths for the cables. Typically, cable lengths are defined between workpoints at the centerline of clevis or shackle pin attachments to the supporting structure (workpoints (2) in the drawing at the top of page 142). Just as fabric patterns are "compensated," cables are fabricated shorter than their final length to compensate for the elongation that they undergo when the roof is prestressed.

A welder affixes cable ear plates to the top of a mast.

Source: This photograph was taken at the shop of Sullivan & Brampton; San Leandro, California.

It is in the nature of conventional steel cables that they must be tensioned to a fairly high level before the individual wires in the cable work themselves into their proper alignment and the cable realizes its full stiffness. In some applications, therefore, the fabricator will "prestretch" the cable to its calculated prestress load, or as much as 65 percent of its ultimate breaking strength, in order to set the wires into their final alignment. This both sets the cable's stiffness to its final "elastic modulus" and allows the fabricator to accurately determine the fabrication length.

The design of the supporting structure must be coordinated with patterning, as well. The geometry and orientation of the ear plates that provide anchorage of cables, for example, must accurately accommodate the angular geometry of cables that is determined only during patterning.

Layout

The contractor must develop a comprehensive installation plan prior to layout and erection of the roof. The plan will consider whether and how the fabric is divided into pieces joined by field seams. The size and configuration of the pieces must address issues such as orientation of the warp fibers, site access, the reach and capacity of available cranes or lifts, expected winds during erection, and cable layouts. (Field seams are logically located at cable lines.)

At the job site, the same care is required in uncrating and laying out the membrane as was taken in preparing it for shipment. On conical roofs, the tension ring at the peak of the structure can facilitate lifting the membrane from the crate, with caution taken to ensure that no creases or other damage are created in the handling of the membrane.

At the Weber Point Events Center in Stockton, California, masts were erected and stabilized prior to installation of the fabric.

In some cases, fabric panels must be joined to each other in the field or attached to cables before the fabric is put into place. If so, the fabric is laid out at grade before erection. A smooth, clean layout surface is required, generally protected by a clean groundcloth or polyfilm, and the fabric must immediately be secured with tag lines so that it is not lifted and damaged by winds. Where feasible, though, the membrane will be lifted from its crate and placed directly over the supporting structural elements, without ever being laid directly on the job-site ground surface.

Erection

Erection of fabric roofs is a highly specialized activity, and, if not performed by a crew experienced with fabric, requires the direction of an experienced superintendent. In general, masts, arches, or other rigid supporting elements are erected prior to the lifting of the fabric membrane itself, with special care taken to ensure the stability of the supports throughout all phases of the erection and tensioning process (see the photographs above and on page 148).

Conical roofs are initially hung from their top ring and attached loosely around the bottom edge. Next, the erectors work their way around the bottom edge, stretching the fabric out in the fill direction and securing it to the perimeter clamping before jacking the top ring up to its final position to bring prestress into the body of the fabric. Suspended roofs may be erected in similar fashion, but in the last step of the process, erectors raise the suspension cables to final elevation instead of jacking the top ring.

Unlike cones, most arch-supported roofs lack a jacking mechanism to bring tension into the roof, and different erection methods

The fabric at the Weber Point Events Center drapes loosely from the upper mast peaks when first lifted into place. The final, unwrinkled, shape is shown on page 25.

are required. As the warp direction of the fabric generally has minimal compensation, the fabric edges perpendicular to the seams can be secured with relative ease before the erectors complete the more labor-intensive work, starting from the center and working out to the two ends, of stretching out the fill fibers and securing the fabric to the edges parallel to the seams (see the photographs on page 149).

The tensioning process is simplified in structures having edge catenary cables instead of clamps. Here, adjusting devices can be installed at the catenary cable anchorages to permit tensioning by pulling only the ends of the cables rather than the entire width of the fabric (see the photograph on page 150).

In roofs employing alternating ridge and valley cables, some or all of the cables may be in place prior to the erection of fabric, which is secured in place to the cable clamping system (see the drawings on pages 102 and 103). Tensioning of the membrane is accomplished by raising the peaks of the roof using one of the mast-jacking mechanisms described in Chapter 6.

Whatever the structural type, the erection procedure must be carefully planned to ensure that the structure remains stable throughout erection. It is nearly impossible to maintain the stability of large, untensioned fabric panels in high winds. In addition to limiting panels to reasonable size, therefore, the fabricator should unpackage the fabric in calm weather and bring it as quickly as prudently possible into its final position and establish tension sufficient to prevent large displacements.

While wind-battered and poorly restrained fabric can readily be damaged, it generally poses only a modest safety hazard. Supporting masts and other elements are greater safety risks, however, and a high standard of care must be taken to ensure their stability through

erection. Catenaries and other cables that are erected as part of the fabric assembly cannot stabilize the supporting elements during erection, so the permanent guy and tie-back cables must sometimes be supplemented by temporary cables. Masts that rest in pots or other devices at their bases in order to allow freedom of rotation rely on the tension in the attached cables of the completed structure to hold these bases in place. If the cables are not in place or are untensioned during erection, however, temporary restrainers may be required at the mast bases in order to prevent updrafts on untensioned fabric from lifting the masts out of the restraining pots.

While the patterning process attempts to develop a membrane that is predictably tensioned throughout its surface, the process is an imperfect one. The designer must be aware that the method of tensioning the fabric determines what adjustment, if any, is available for correcting small deviations from exact patterning that occur in the field due to variations in material properties or imprecision in the analysis. Structures that are tensioned by pulling the boundary of the fabric to a fixed edge generally allow no adjustment, while mast-tensioning devices provide adjustment at the peak of the tent. Catenary, valley, or ridge cable adjustments allow the membrane to be tensioned along the length of the cable, thereby providing more flexibility in the application of prestress.

During erection, the fabric in the vicinity of mast peaks and other singularities must be checked periodically for its level of stress. Wrinkling provides an easily recognized indication that an area of fabric has little or no tension in the direction perpendicular to the wrinkles, while overstress may be more difficult to spot. Workers may "test" the membrane's stiffness under foot or hand pressure, as the stiffness will increase with prestress level. High curvature,

Workers stretch fabric into position with come-alongs [BELOW] at Baypointe Station before securing the edges with clamp bars [ABOVE].

Workmen at Weber Point Events Center pull the corners of the membrane outward and secure them to the perimeter masts using temporary cable and "come-along" type winches. Tensioning is completed by cinching down the anchor bolts to the foundation (see the right side of the photograph).

short fabric spans, and the proximity of supports can all make the fabric feel stiffer, however, so that a "hard" surface is not necessarily indicative of excessive prestress.

Erectors sometimes make use of mechanical tension measuring devices. Birdair, for example, uses a device designed by engineer Jim Ford that measures the force required to push a chisel-shaped probe down into the fabric until the edge of the cylinder surrounding the probe lies flat against the fabric. By orienting the chisel edge perpendicular to the warp and fill fibers in turn and correlating the pressure with data established on test samples, the erector can obtain a good estimate of the fabric stress about both axes at any location on the membrane (Ford 1999, personal communication).

In general, structures with simple, slowly varying curvatures provide the least demand for accurate patterning, while sharper curves or changes in curvature provide a higher demand. The corners of arched roofs and the peaks of cones are most prone to wrinkling or overstress. At mast peaks or junctures with perimeter posts, therefore, turnbuckles are sometimes supplied to provide localized "tuning" of fabric after the overall tensioning is completed. If necessary, the final position of mast- or cable-tensioning mechanisms can be varied from the theoretical position.

Irregularities apparent during tensioning may be the consequence of patterning inaccuracies, but they may also be indicative of inaccuracy in the location of the workpoints in the supporting structure. In any event, they must be noted and corrected promptly, as small irregularities noted early in the erection tend to become more significant as tensioning proceeds and may lead to problems of appearance or performance in service.

Final tensioning of a fabric roof is analogous to the tuning of a guitar. Musicians recognize that adjustment of one string causes deflections in the guitar neck that change the tension (and pitch) of each of the other strings. They adjust the strings incrementally, and recheck and retune each in turn until a final harmony is achieved. Similarly, tension must be provided gradually throughout a fabric structure and, where multiple tensioning points are used, the erectors apply tension incrementally, moving repeatedly in sequence from point to point. Final tensioning has the goals of developing the desired stress field in the fabric and removing any wrinkles.

Adjustable turnbuckles, cable terminations, drawbolts, and masts all provide means for correcting tensioning problems. Adjusting devices are expensive both in terms of fabrication cost and visual clutter, however, and they create the hazard that inexpert erectors will create new and unanticipated tensioning problems in the effort to correct initial wrinkling. In developing the details for a fabric structure, the designer should endeavor to provide adjustment mechanisms that are sufficient to tune the membrane but that are otherwise minimized (Daugherty 1999).

Before erection can be considered complete and the structure turned over to its owner, complete inspection is required, and dirt or stains that may have adhered to the fabric during fabrication and erection should be removed (see photograph on page 152). Inspection may include subjective evaluation of tension levels throughout the membrane. The field of the fabric should be checked for tears or other blemishes that may have occurred during the erection. Mast peaks and other discontinuities require particularly close inspection for wrinkles, small tears, or protruding edges from fabric clamps or other devices that might cause future damage to the fabric under movement.

Coordination with Other Phases of Construction

It is an unusual and perhaps fortunate fabric roof contractor who also acts as general contractor, with responsibility for both the fabric

Water jets may be used for general cleaning at the end of erection.

membrane and the supporting structure, foundations, and other interfaces. More often, he installs the membrane, cables, and rigid structural elements directly above or below the membrane, and he must coordinate these elements with a supporting structure or foundation and various additional elements that are installed by others. These additional elements may include gutters, window wall, roofing mechanical systems, lighting, and fire sprinklers. Coordinating them with the fabric installation is critical to the success of the project.

Initial coordination with other elements of construction occurs during design, with careful consideration of geometric interface with structural supports as well as loading of and possible deflection of the supports. Similar consideration is required for any nonstructural interface elements.

The fabric roof erector also must be closely observant of interface conditions in the field prior to beginning erection. As a minimum, he must visually confirm the size and placement of anchorage and other interface elements. While the general contractor or other subcontractors may be legally responsible for the accuracy of placement in support structure interfaces, it may nonetheless be necessary to obtain an accurate survey of these conditions in some structures. In general, accuracy of anchorage placement is more likely to be an issue in concrete construction than in steel, and accuracy may vary with the sophistication of local construction practices. In one project that we completed in a developing country, we learned upon delivery of the fabric membrane to the site that errors had been made in the placement of concrete foundations. One anchorage deviated from its design location by 1.8 m! Our determination of cable lengths to the millimeter was meaningless in the face of such conditions. Validation of support conditions prior to erecting fabric can be critical to the success of the project.

As discussed earlier in this chapter, many structures require full layout of the membrane for attachment of cables and clamps prior to erection. Coordination of this requirement with the space constraints of the site and the needs of other trades is best done as part of contract negotiations. The layout site must be properly situated to allow smooth transference of the membrane onto the supporting structure. From the time of layout until the work of all trades has been completed, the fabric contractor must take whatever measures are necessary to ensure that the fabric is not exposed to soiling agents or any object that might cut or abrade it. Paint can pose a particular hazard. During erection of one of my recent designs, the fabric was pulled across a steel supporting structure that had just received touch-up paint. The dried paint has proven difficult to remove and led to the fabric supplier's threat of voiding its warranty.

While fabric roofs are lightweight in comparison to other roof structures, cranes or other equipment must lift them in large segments and often require long reaches. Like interface and layout requirements, lifting equipment requirements must be anticipated early and coordinated with other trades.

Maintenance

Properly designed and constructed fabric roofs generally require little maintenance until such time as degradation from ultraviolet radiation or other sources necessitates replacement of the fabric. Owners should be supplied with kits for repair of small tears. A complete repair kit will include round and rectangular patches of various sizes; a larger piece of rolled fabric goods; material and tools as appropriate for installing patches of a particular material, along with instructions on their use; safety guidelines; and references for whom to contact in the event of serious problems. The owner may require the services of the roof supplier to effect patching or replacement of sections of fabric where more severe damage has occurred. Periodic inspection by the manufacturer or other qualified personnel is recommended.

Periodic cleaning of fabrics may be required, depending both on the fabric used and the environmental exposure. At the University of La Verne Campus Center, for example, the polluted San Fernando Valley air causes regular soiling of the PTFE-coated fiberglass fabric. By informal agreement with the local fire department, the roof is washed bright white periodically with a fire hose.

Other fabrics are less forgiving of soiling. PVC-coated fabrics must be regularly cleaned to avoid loss of translucency and unattractive discoloration. The fabric may be embrittled by repeated contact with soaps, solvents, or oils, however, and the fabrics should be cleaned only in accordance with the manufacturer's instructions, using soft brushes, water, and mild detergent. PVC-coated fabrics also may discolor with long exposure to the sun's ultraviolet radiation, although PVF or PVDF topcoatings may reduce this effect (see Chapter 4).

Polyester-based fabrics also have some tendency to creep (stretch) under sustained prestress load, and inspection by the manufacturer several months after installation is advised in order to determine if retensioning of the membrane is appropriate. In designing certain connections, the engineer and installer must account for the possibility of retensioning (or of removing and replacing a worn-out or damaged membrane). Where temporary jacking frames or similar devices are used, for example, long rods must be left in place or provision made for coupling them in order to permit re-installation of the frame for rejacking.

8 Nonstructural Performance Parameters

In a conventional roof assembly, structure is one of many elements sandwiched between roofing on the outside and a ceiling on the inside. Others include insulation, lighting, and components of the building's heating, ventilation, air-conditioning, plumbing, fire safety, and electrical systems. In a fabric roof, however, insulation and at least a portion of the lighting are products of the fabric itself (though perhaps a double-layered fabric), and the remaining elements of the conventional roof assembly (including supplementary artificial lighting) are either exposed beneath the fabric or divorced entirely from the roof construction.

Fabric roofs are, first of all, structures, and the focus of the book thus far has been on the structural properties of materials, structure as a form giver, structural analysis and details, and fabrication and erection of the structure. A successful fabric roof must carefully address several aspects of performance other than structure, however, including lighting, energy, acoustics, and fire safety. These items are considered in the pages that follow.

Daylighting

One of the most exciting properties of fabric roofs is their ability to transmit light in a manner that creates vividly lit interior spaces during the day and luminously glowing exterior forms at night. While light behavior is a complex function of the interplay between various properties of both the fabric and the light source, the critical lighting property of structural fabrics is translucence: the ability of the fabric to transmit a portion of the light that strikes it directly to the space beyond. Light transmission properties of fabrics vary widely from the near opacity of a stage backdrop to the near transparency of delicate hosiery.

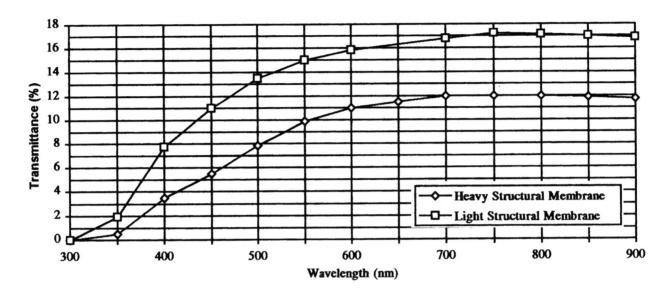

Spectral transmittance of PTFE-coated fiberglass fabrics varies across the spectrum from ultraviolet (wavelength less than 400 nm) to infrared (wavelength greater than 700 nm). Data shown is based on Chemfab's Sheerfill I heavy structural membrane and Sheerfill V light structural membrane.

Source: Birdair, Inc.; used with permission.

Light Intensity and Quality

The translucence of contemporary structural fabrics like polyvinyl chloride (PVC)-coated polyester and polytetrafluoroethylene (PTFE)-coated fiberglass generally ranges from 5 to 25 percent. The resulting potential for daylighting is tremendous and has made fabric popular for sports facilities, exhibit halls, and atriums or other sky-light-type applications.

An exterior light level of 75,000 lux, typical on a sunny day, results in an interior light level of 11,300 lux using a fabric of 15 percent translucency. Typical modern office light levels, by contrast, are about 1,000 lux. This lighting level has proven ample for lush growth of certain plants indoors. It has not been sufficient for grass growth, however, so football and baseball stadiums require the use of higher translucency products recently on the market if they are to avoid the use of artificial turf.

Typically, fabrics do not transmit light evenly across the color spectrum. A lightweight PTFE-coated fiberglass material, for example, may transmit nearly 12 percent of visible red or infrared light, but less than 4 percent of violet light and only a tiny portion of ultraviolet light (see the chart above) (Birdair 1996). This characteristic helps to protect interior finishes and building contents from fading, but it may limit the variety of plant life that will thrive inside (Sinofsky 1985).

Most structures have exploited the translucency of fabrics to reduce artificial lighting requirements. Structural fabrics typically have the additional property of diffusing transmitted light so that it scatters in all directions, lending an even and often shadowless quality to the light that is not unlike that produced by strong sunlight passing through scattered clouds or light fog. The light is soft and flattering. In speaking of the Bullock's department store in San Mateo, California, former Bullock's of Northern California chairman Paul Heidrich noted, "For the first time, a woman at the cosmetics counter will be able to see herself in natural light" (*San Jose Mercury* 1981). At the other fabric-roofed Bullock's store, in San Jose, California, store managers found that light passing through the fabric illuminated the merchandise under it so attractively that sales volume in the departments under fabric was 10 to 15 percent higher than in comparable departments in its other stores (*Business Week* 1979). The lack of shadow may hinder the recognition of shapes, however, and artificial lights may be needed to accent selected products during daylight hours, a common retailing requirement. Where heavy spotlighting is required or a space must be darkened during daylight, however, translucent fabrics may not be appropriate.

The magnitude of daylighting is often altered by varying the translucency of the fabric or by adding a liner membrane or insulation. Liner membranes—which are incorporated into structures to improve insulating value, soften acoustics, prevent vapor condensation, and improve fire resistance—are typically manufactured with translucence between 20 and 30 percent. The combined lighting properties of the two membranes can be considered in accordance with the following formula:

$$P = [P_1 \ P_2 + P_1 P_2 R_1 R_2 + P_1 P_2 (R_1 R_2)^2 \ ...]$$
$$P = (P_1 \ P_2)/(1 - R_1 \ R_2)$$

where

$P =$ Light transmission of the assembly
$P_1, P_2 =$ Light transmission of the individual fabrics
$R_1, R_2 =$ Reflectance of the individual fabrics

Combining a typical structural membrane of 75-percent reflectance and 15-percent transmission with a typical liner membrane of 75-percent reflectance and 20 percent transmission yields an assembly with light transmission of 6.9 percent. As the example demonstrates, fabric reflectance substantially increases light transmission of the assembly by permitting light reflected off the liner to be reflected back off the structural membrane and into the space. Changing reflectance to zero without changing transmission of the fabrics in the above example reduces light transmission to 3 percent.

Illumination of Roof Forms

The light properties of liner membranes also impact the ability of someone inside the building to "read" the roof form. The form of a single-membrane roof is usually well delineated by supporting mem-

Light below the structural and liner membranes of the Denver Airport terminal is at the same time bright and diffuse.

Photograph copyright Robert Reck.

bers, cables, and fabric seam lines, whose opacity leaves them as dark curves that follow the fabric shape. By adding a liner, however, light transmitted by the structural membrane is diffused by the liner to erase all shadows on the structural membrane. What remains is a warmly glowing luminous ceiling (see the photograph above) (Ishii 1995).

When desired, the interior readability of the roof shape can be improved by the use of supplementary artificial lighting that brightens portions of the roof facing the artificial lighting while leaving other areas darker. This effect is generally maximized by introducing the lights at a shallow angle to the membrane surface.

Nighttime lighting effects inside fabric structures are dramatically different from those in daytime. The natural luminosity of the roof under daylight is of course lost, although it may be partially regained through the use of interior light sources that are directed at the roof to take advantage of the generally high reflectivity of architectural fabrics (see the photograph at the top of page 159). As noted

Reflected light from spotlights below creates dramatic nighttime effects on the Hollywood & Highland fabric canopies.

Source: Don Douglas; used with permission.

earlier, sunlight takes advantage of the reduced translucency at seam lines to distinguish the roof form in the daytime, an effect lost with nighttime interior lighting. Lights directed at shallow angles to the membrane are therefore particularly useful in varying the brightness of the fabric surface in a manner that distinguishes the roof form. Viewed from the outside, nighttime interior lights directed at the membrane invert the effects of daylight on the interior of the structure by creating a luminous surface that is distinguished by the dark lines of seams, cables, and supporting structure.

A summary of the characteristics of various roofing assemblies is given in the table on page 160 (Huntington 1987).

Effects of Dirt Accumulation

Critical to maintaining daylighting over time is the cleanability of the fabric, a property of materials and topcoats that is discussed in Chapter 4.

| | Assembly Number | | | | | |
Properties	1	2	3	4	5	6
Solar properties (percent)						
Reflectance	10–50	30–75	70–75	70–75	25	70–75
Absorption	50–90	13–68	12–18	19–26	75	20–28
Transmission	0	2–12	7–18	4–6	0	2–5
U-Value						
Summer (3.4 m/sec wind)	Varies	0.75	0.81	0.45	0.18	0.08–0.14
Winter (5.7 m/sec wind)	Varies	1.15	1.20	0.54	0.18	0.08–0.14

Notes: Assembly 1, conventional roofing; Assembly 2, PVC-coated polyester fabric; Assembly 3, PTFE or silicone-coated fiberglass fabric; Assembly 4, fiberglass fabric with liner and 250 mm air space; Assembly 5, LaVerne (fiberglass fabric with opaque insulation); Assembly 6, fiberglass fabric with translucent insulation.

Source: Huntington 1987.

Solar and energy properties vary widely for different fabric assemblies.

The mechanics of dirt deposition and roof shaping and details also play an important role in limiting the effects of dirt accumulation on light transmission and appearance. In general, dirt first accumulates on flat surfaces near the top of the roof, then is washed down the sloping sides of the roof by rainfall. As the rainwater evaporates, streaky dirt deposits are left along the sloping sides of the roof. Devices such as gutters, which are used to capture water at the edges of conventional roofs, prevent water from running down the walls of these buildings and leaving streaks. Gutters and other rain diverters have seldom been used in fabric roofs, however (Ishii 1995).

Energy Use

The fundamental purpose of a tensioned fabric structure, like that of any roof, is to provide shelter. Yet, with few exceptions, fabric structures are designed with limited knowledge of, and little attention to, the effects of design decisions on the building's energy behavior. It is interesting to compare this to the high level of care that is now generally taken in form finding and analysis, the structural parameters of roof performance. The difference may be one of expediency. We have seen in previous chapters that the structural engineer of a fabric roof cannot add steel reinforcement to compensate for inaccurate analysis in the manner that the engineer of a concrete roof can—the inaccuracy will be reliably manifested by wrinkled or torn fabric. In designing fabric roofs for energy efficiency, however, engineers can (and often do) compensate for poorly conceived designs by "reinforcing" the capacity of heating and air-conditioning systems.

As fabric roofs are used for more sophisticated applications, the crudity of past approaches has become clearly unacceptable. The typical occupant of a fabric-roofed building has no understanding of or interest in the stress level or form finding of the roof membrane. He is acutely aware of relatively subtle gradations of hot, cold, wet, or dry, however, and the knowledge and attention of designers should reflect this reality.

Fabric Thermal and Lighting Properties

The design of energy-efficient fabric roofs must begin with a general understanding of the thermal and lighting properties of fabric. Fabrics in common use are characterized by low insulating ability and low thermal mass. Because of this, the thermal resistance of uninsulated fabric roofs depends almost completely on the heat transfer mechanisms occurring at the fabric surfaces, termed "film resistances." Heat is transferred from surfaces by convection to the air moving over its surface and by longwave (infrared) radiation to other surfaces or to the sky. The thermal resistivity or "R" value provided by these mechanisms is low, and the energy behavior of fabric roofs is therefore highly dependent on their optical behavior, which is characterized by high reflectivity, low absorption, and a range of translucency (Sinofsky 1985). These characteristics have made them readily applicable to use in temperate or hot climates with high solar radiation. In these conditions, the low insulating value does not result in high heat loads, the reflectivity reduces heat gain, and the translucency can be utilized for natural daylighting to reduce lighting cost.

The interior daylighting provided by architectural fabrics can slice artificial lighting requirements sharply. Similarly, the high reflectivity of fabrics (especially white) can bring impressive reductions in air-conditioning load. As much as 90 percent of the energy striking a conventional roof is absorbed into the roof, and the remaining 10 percent is reflected. Tremendous heat is stored in the roof, and much of it subsequently radiates into the space. White glass fiber fabrics, by contrast, absorb between 8 and 21 percent of the light (Chemical Fabrics Corp. 1996) and consequently store and radiate much less heat into the space. Absorption may be further reduced by alterations in conventional material chemistry, such as the addition of reflective titanium oxide particles to PTFE coatings (Beitin 1982). This addition also reduces heat gain in applications with high air-conditioning load.

A more nuanced understanding of fabric roof energy behavior, however, requires consideration not only of the overall parameters of heat transfer, but also of the variations in temperature throughout the space. Most conventional buildings have substantial thermal mass, so temperatures both within the space and of interior surfaces are relatively constant throughout the building (Harvie 1995).

Thin fabric membranes can experience rapid changes in temperature in response to changes in external temperature or sunlight,

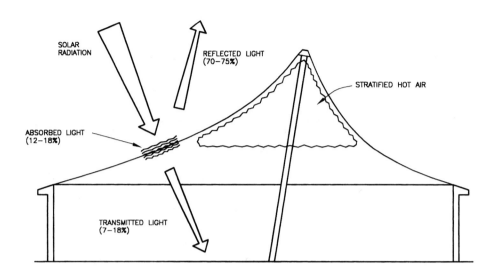

Fabric's solar properties combine with heat stratification to give good energy performance in warm climates.

however, and temperatures within the space can therefore vary substantially, depending on proximity to surfaces of different temperatures. On sunny days, the fabric roof surface temperature will be substantially higher than that of the floor below. This factor, combined with the tendency for hot air to rise and stratify directly under the fabric, well above the heads of building occupants (see the drawing above), results in a temperature drop of 5°C or more between the upper and lower elevations within the space. This behavior may result in a surprisingly high level of comfort inside fabric-roofed buildings in sunny, hot climates, despite the low insulating value of the fabric (Huntington 1987).

In cold weather, the low thermal mass and insulating values of fabric create a temperature gradation across the height of the space that works to the disadvantage of fabric. This is because heated air rises to the top of the building, where uninsulated fabric allows it to escape. However, the energy use characteristics of the building can be substantially altered by the use of a lightweight inner "liner" fabric membrane to increase the insulating value of the fabric. Used with a 260-mm air space, a liner decreases the U value for winter conditions with a 6.7-m/sec wind from 1.20 to 0.54 (see the photograph on page 160).

In cold climates, attainment of U values low enough to meet the requirements of some recent energy codes requires the use of insulation, however. The University of La Verne Campus Center uses 38-mm-thick glass fiber metal building insulation that yields a U value of 0.18 (R-5) but that is completely opaque (see the photograph at the top of page 163). At La Verne, the inability to capitalize on the energy savings and occupant comfort that a translucent fabric can provide offsets the insulating value.

More recently developed glass fiber products, originally developed for use in furnace filters, provide good insulation while retaining substantial translucence. The Lindsay Park Sports Centre in Calgary, Alberta, built by Birdair in accordance with the design of Chandler Kennedy Architectural Group and engineer Geiger Berger, has an outside fabric of 18 percent translucence and 400 mm of insulation with 20 percent translucence. The insulation is supported on an open webbing material, with a Tedlar® film of 90 percent translucence suspended below. The assembly (see the photographs on page 40) boasts R-12 insulation and 4 percent translucence, demonstrating the ability of fabric roofs to be energy savers on all counts (Chemical Fabrics Corp. 1984).

Loss of the energy savings and visual delight of daylighting mitigate the value of the opaque insulation at University of La Verne.

Condensation

In the colder climates where insulation may be used, measures must often be taken to prevent excessive condensation, particularly for applications such as swimming pools, zoos, or botanical gardens that may have damp interior environments. Condensation is likely to occur when the temperature of the membrane and the relative humidity of the inside air are such that the air on the inside surface of the membrane can reach the dew point. Where this is a possibility, consideration should be given to venting the inside air, installing condensate gutters, or providing an air circulation system.

The decision to use a liner membrane will affect condensation behavior. Lightweight liners are generally permeable to water vapor, and condensation can therefore form on the inside surface of the outer structural membrane. The same liner materials are generally impervious to water as a liquid, however, so that dripping condensate is generally captured by the liner and brought to the edge of the membrane without falling into the interior space.

Condensation can be a particular problem on insulated roofs, where it may soak and limit the effectiveness of insulation. Such problems (eventually resolved) occurred on the roof of the East Area Health Center in Detroit, Michigan, installed in 1983, where 100-mm fiberglass batt insulation below the structural membrane was soaked by condensation. "In the north," says Kent Hubbell, professor of architecture at Cornell University, "you have to have the right kind of vapor barrier in the right place, you have to ventilate the space between the vapor barrier and the outer membrane, and you have to provide for prompt drainage of any condensation that does occur" (Gorman 1992).

Heated ventilation air can be blown into the space between the structural membrane and insulation in order to raise the temperature of the air space and control condensation. Care must be taken, however, to maintain the cleanliness of the ventilation air, so that the insulation is not soiled to cause a dingy appearance and a reduction in translucency.

In climates that combine warmth with high humidity, caution must be taken against the growth of mold or algae caused either by condensation or standing water on the outside of the fabric. While such growths may be removed by cleaning, it is preferable to prevent the formation of condensation in the first place, using measures as described above, and to provide adequate roof slope to drain.

Measuring Energy Performance

To summarize the energy behavior of fabric roofs, translucency provides daylighting that reduces artificial lighting requirements substantially, reflectivity prevents the storage and radiation of heat in hot weather, and liners or insulation ensure retention of heated air in cold weather. Both liners and insulation reduce the translucence of the fabric assembly and the amount of daylighting that can be provided. Studies have shown, however, that translucence in excess of that required for appropriate daylighting levels is usually detrimental to overall energy use. While this "optimum" translucence varies with building use and design, it typically occurs at about 4 percent, well below the translucence of structural fabrics, such that substantial insulation is generally beneficial (Sinofsky 1985).

Accurate prediction of a fabric-roofed building's energy behavior is at present available only through the use of relatively arcane techniques of computational fluid dynamics. Using this methodology, the points within a structure are analytically linked so that the effect of change in the state of one location on another location can be predicted in a manner analogous to the way the structural engineer's finite element analysis can predict the effect of a load at one location on shape and stress throughout the membrane.

Given the number of variables involved, there is obviously no general answer to the question of whether a fabric-roofed building is more energy efficient than one with a conventional opaque, insulated roof. A broad comparison can be made, however, by considering the balance between fabric's advantageous daylighting and disadvantageous heating loss in a particular climate (Beitin 1982). A structure may be considered to have a calculable number of "degree hours" below a certain base ambient temperature in a given locale. Furthermore, the conventional structure may be expected to use a certain wattage of artificial lighting (per unit of floor area) during certain assumed hours of use. The relationship between the conventional roof's artificial lighting burden and the fabric roof's heating burden can then be graphed to approximate their relative energy efficiency.

Substantial data is now available to verify the thermal performance of fabric-roofed structures. In 1980, the U.S. Department of Energy sponsored a computer modeling study of the Stephen C. O'Connell Center fabric roof in Gainesville, Florida (Basjanac 1980). This study projected an energy use of 15.8 W/m^2 for the fabric structure, as compared to 20.6 W/m^2 for a conventional roof. The warm climate of Florida, of course, assisted in achieving this energy advantage.

Another study investigated the Bullock's department store in San Jose, California, a structure with a 1,600 m^2 fabric skylight in the middle of a conventional roof (Oberdick et al. 1981). In spite of the fact that fans under the skylight provide conditioned air for most of the building, the study indicated that energy use under the fabric skylight was about 15 percent less than under the conventional roof.

The same researchers analyzed the energy consumption for the Bullock's roof (assuming a full fabric liner instead of the partial one in the real structure) in the colder climate of Mason City, Iowa. This analysis still indicated superior performance under a fabric roof: 38.7 W/m^2 as opposed to 48.2 W/m^2 under the conventional roof.

La Verne, with its opaque insulation, has energy performance closer to that of a conventional building than a structure like Bullock's. Review of college records (Huntington 1986) indicates total energy consumption of about 20.9 W/m^2, a fairly low value even for Southern California's warm and sunny climate. Predictably, energy use is higher in the hot summer months than in the mild winters. La Verne uses approximately 24.2 W/m^2 during the months of July through September and 18.8 W/m^2 during January through March.

Adverse Weather Conditions

Fabric roof profiles must be configured to prevent ponding of all forms of precipitation, including rain, snow, and ice. Avoidance of ponding is critical, as the weight of accumulated water will cause deflections that allow greater ponding, leading to a potential overloading condition. In general, a minimum slope of 4° to 5° should be provided.

Moderate snowfall can successfully be resisted in structures that have prestress and slope sufficient to prevent deflections large enough to lead to ponding, additional deflection, and eventual overload of the roof. Relatively high roof slopes are also useful in helping the slippery fabric surface shed snow.

Snow-melting equipment, usually in the form of a forced air furnace and fans to circulate heated air between the structural and liner membranes, has been used in regions of heavy snow to prevent buildup on the roof. Because the snow load capacity of an air-supported roof is limited to the difference between internal air pressure and dead load (usually about 200 N/m^2), effective snow melt systems are critically important in air-supported roof design.

The effectiveness of such measures is limited by the reliability of both the equipment and its operators, however, and the well-publicized deflations of air-supported stadium roofs have demonstrated the pitfalls of this approach even with relatively sophisticated designs and operators. It is important to note that these systems do not effectively remove snow when the roof is heated after buildup has already occurred.

When snow-melting systems employing heated air are used, condensation must be either prevented with appropriate ventilation or collected. In single-membrane roofs, condensation gutters may be provided along cable lines. In double membranes, the heated air is blown between the two layers of fabric, and the condensation that develops on the underside of the structural membrane will drip onto the liner, which should be designed to carry water to the perimeter of the structure where provision must be made for its collection.

Lightning may be a hazard in fabric roofs, and many structures require the installation of lightning arrestors at roof peaks. The arrestors may be attached to ridge cables to provide grounding. In structures with cable nets, arrestors may not be required (Geiger 1989).

Acoustics

The acoustical performance of fabric roofs is generally measured by their reverberation characteristics and ability to resist the transmission of noise. Reverberation is critically important in venues featuring musical performances or public speakers, while reduction in sound transmission is vital either when outside noise must be kept out (as at an airport) or when the outside must be shielded from loud music or noise inside the space.

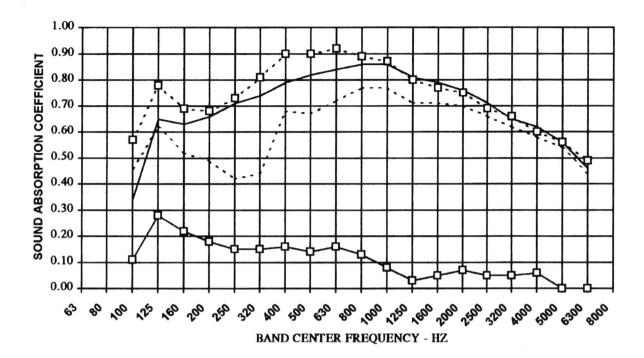

In considering reverberation, the designer must evaluate the reverberation time inside the enclosed space, which is directly proportional to room volume and inversely proportional to the sound absorption of the fabric and other materials in the space (Birdair 1996). Intelligibility of speech, in particular, requires relatively short reverberation times, which are easily realized in small enclosures but more difficult to attain in large sports or entertainment facilities. In these uses, high absorption can be critical to successful acoustics. The aggregate sound absorption of a material over a wide range of frequencies is measured by its "noise reduction coefficient" (Moulder and Merrill 1982), which may vary from 0.10 for a heavy structural membrane to 0.80 for a lightweight secondary membrane (see the chart above). The acoustic properties of such fabric liners (along with their energy, lighting, and fire resistance effects) are critical to the success of certain roofs.

The chart above also illustrates the considerable variation in absorption that fabrics have at different frequencies. Structural weight membranes have high reflectivity of sound (as reflected in low absorption) at frequencies above 250 Hz (middle C, for reference, is 256 Hz). This characteristic can be useful in performances of acoustic music, where the reflectivity helps the musicians hear themselves, but it can detract from amplified performances by reflecting too "bright" a sound to the audience (Rebeck 1991).

Like absorption, sound transmission loss (measured in decibels) as sound passes through a given material varies widely across the frequency range (see the chart on page 168), and is quantified in aggregate by its "sound transmission coefficient" (STC) (Moulder and Merrill 1982).

Single structural weight fabric membranes offer a poor barrier to

NOISE REDUCTION COEFFICIENTS (NRC)
─□─ Heavy Structural Membrane (NRC=0.10)
····· Heavy Structural Membrane + Liner (NRC=0.65)
──── Liner (NRC=0.80)
-·□·- Heavy Structural Membrane + Liner & 50mm Insulation (NRC=0.80)

Sound absorption varies across the audible range. The noise reduction coefficient, a measure of the aggregate absorption across the sound spectrum, is improved by the use of liners and, to a lesser degree, insulation. Data shown is based on Owens-Corning Fiberglas Structo-Fab 120 liner and Structo-Fab 450 structural membrane.

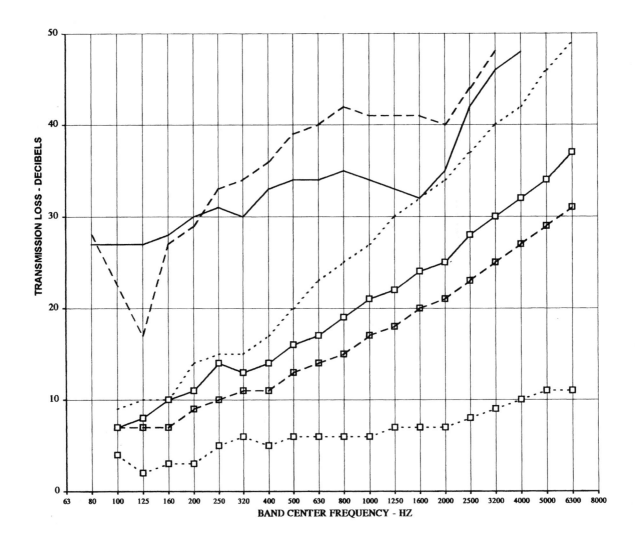

SOUND TRANSMISSION COEFFICIENT
- ⋯□⋯ Liner (STC 7)
- —⊟— Heavy Structural Membrane (STC 17)
- —□— Heavy Structural Membrane Plus Liner (STC 21)
- ⋯⋯⋯ Heavy Structural Membrane Plus Liner & 50mm Thick Insulation (STC 25)
- ——— 3/8" Laminated Glass (STC 36)
- — — — 1-3/16" Insulating Laminated Glass (STC 40)

Sound transmission loss also varies widely across the audible range. The sound transmission coefficient represents an aggregate of transmission loss across the sound spectrum. Data shown is based on Owens-Corning Fiberglas Structo-Fab 120 liner and Structo-Fab 450 structural membrane.

the transmission of objectionable sounds, (particularly those in the low-frequency range), providing only STC 17. The addition of a liner yields a modest improvement in resistance to sound transmission (to STC 21). The addition of fiberglass insulation results in further improvement (to STC 25). Even with these improvements, however, fabric roofs cannot compete with conventional roofing assemblies in sound transmission loss, as given in the following examples (Birdair 1996):

1. Wood or steel roof deck with insulation and built-up roofing (STC 30)
2. 10 mm laminated glass (STC 36)
3. 30 mm insulating laminated glass (STC 40)
4. Wood or steel roof deck with gypsum board ceiling (STC 45-50)

As described above, sound reflectivity can be decreased (by increased absorption) and transmission loss increased by the installation of lightweight, porous liner fabrics. Fiberglass insulation

The domed shape of air-supported roofs, like the Pontiac Silverdome in Michigan, tends to focus sound reverberation toward the middle of the field below, an effect that is partially mitigated by the use of suspended acoustical baffles.

Source: Birdair, Inc.; used with permission.

between the two fabric layers can further increase transmission loss. The effects of such measures on daylighting, insulation, and fire safety must be considered, however.

The slope of a fabric roof also can affect its acoustic performance. The characteristic anticlastic curvature of a saddle or tent-shaped roof tends to diffuse and blend sounds, while a synclastic air-supported or cable dome roof may focus sound in the manner of a lens. Some air roofs have utilized quilted baffles to increase absorption (see the photograph above) and reduce the highly reverberant roar that these roofs may be subject to.

Greater understanding of the acoustic behavior of fabrics and improved skill in the manipulation of shapes and material assemblies have made it possible to employ fabric in the most demanding of musical applications. One of the most sophisticated of these roofs is the Carlos Moseley Pavilion, a deployable, truck-mounted canopy that is set up at varying locations throughout the New York City metropolitan area for use by the New York Philharmonic Orchestra and Metropolitan Opera (see the photograph on page 170).

Fire Safety

On July 6, 1944, 168 people burned to death in the Ringling Brothers, Barnum and Bailey big top fire, the worst disaster in the history of tent construction (Hatton 1979). Our notions about the fire safety of fabric structures are colored by the memory of this and other, less spectacular incidents. Contemporary architectural fabrics have been formulated to avoid such disastrous fire behavior. The test parameters by which the fire resistance of materials is gauged, and the general performance of different fabrics on these tests, are discussed in

The Carlos Moseley Pavilion provides both a strong architectural statement and an acoustically effective stage backdrop.

Copyright Jeff Goldberg/Esto.

Chapter 4. Fire safety in fabric-roofed buildings is not determined by fabric material properties alone, however, but must also consider the building's use, the location of the fabric, the fire-resistant properties of other materials used, and the use of sprinklers or other fire-suppression measures.

There are specific differences in the requirements of the United States code bodies, but their fire-resistance requirements are similar. For purposes of the *Uniform Building Code* (2000), materials that can both pass ASTM E136 and have flame spread (per ASTM E84) of less than 50 qualify as noncombustible and can be used in Type II-N construction. Combustible roofing materials are restricted to use in Type V-N construction. Additionally, noncombustible roofs must provide resistance to exterior fire exposure in accordance with ASTM-E108 that is consistent with the building occupancy. Noncombustible materials used exclusively as a roof and located more than 7.5 m above any floor, balcony, or gallery are deemed to comply with the roof construction requirements for Type I and Type II fire-

Suspending sprinkler pipes directly from the roof's ridge cables provides a simple means of support while limiting the visual obtrusiveness of the fire suppression system at Bullock's department store in San Mateo, California.

resistant construction. The same height minimum is applicable to structural members supporting the membrane. Available architectural fabrics do not provide for isolation of fire inside the structure and are therefore not used as "rated" (1, 2, or 4 hour) assemblies.

A weakness of each of the United States model codes as a standard for fire resistive design is that they were written with the characteristics and fire performance of conventional building materials in mind, and they do not address the special requirements of tensioned fabric roofs. At least one recent model code, ASCE Standard 17-96, "Air Supported Structures" (ASCE 1996), was written to address this deficiency. This document provides a fire rating "Class" (1A to C) based on the fabric's performance on the four standard tests listed above, and places limitations on the occupancy of buildings of each Class.

Many fabric-roofed structures have incorporated sprinklers as an automatic fire-suppression system. Incorporating them into a fabric roof without damage to its appearance can be difficult due to exposed, curving roof forms and widely spaced structural members. In addition, the piping must be designed to provide sufficient articulation to accommodate movement in the flexible roof membrane. Well-designed systems have performed admirably; for example, the sprinklers at La Verne have not leaked or malfunctioned in 30 years of operation.

Often, sprinkler lines are installed to follow the curves of arch or cable lines so as not to "read" as additional structural elements (see the photograph on page 171). Arches or cables may be spaced 15 m or more on center, however, while standards such as NFPA 701 generally require a sprinkler head for every 15.5 m^2 of floor area, with no head more than 3.0 m from an adjacent head. Some structures have employed side throw sprinklers at the perimeter to cover the space, while in others, designers have been granted code variances to position sprinklers only along the structural members.

Some applications also have required gravity or powered smoke vents to exhaust smoke accumulation from a fire. Vents of appropriate size may be provided at the peaks of cones or the apexes of arches, or they may be incorporated into the conventional construction around the perimeter.

9 The Contemporary Fabric Structures Industry

Because the contemporary practice of architecture is more business than art, the unconventional organization of the design and construction team for tensioned fabric structures has limited their acceptance among architects. Most buildings today are designed by a team of engineering specialists working under the direction of an architectural generalist. Their design, once completed, is then constructed by a contractor. The architect and his consultants oversee the work of the contractor to ensure that the building is built in general accordance with their design.

These relationships are explored in greater detail later in this chapter. Central to this model, though, is the architect's primacy in the design process. Decisions about the form of the building are made by the architect, though he may utilize the feedback of the structural engineer, and, to a lesser extent, his other consultants. The shapes of tensioned fabric roofs, though, like most bridges or concrete shells, are governed largely by considerations of structure, and the architect relies on the input of a structural engineer with specialized expertise in tensioned fabric roofs to find the shape of the roof. Many structural details, furthermore, are left exposed in the finished structure, and the architect relies on the structural engineer to design them to be visually pleasing and appropriate to his design concept.

For the architect who wants to maintain nearly complete control over his design and who does not want to be at the mercy of a highly specialized and difficult to replace consultant, the nature of fabric structure design practice may dissuade him from entering the field. In the late nineteenth and early twentieth centuries, the structural design of bridges, high-rise buildings, and long-span roofs represented some of the world's most advanced technology, and some of the most visually exciting, as well. A romance and excitement surrounded the leading structural engineers that can be compared to our more modern infatuation with computer innovators like Steve Jobs. Many architects sought to collaborate with the great structural engineers and to share with them in the application of new technology to the creation of beauty. Together, they created a wave of great bridges, long-span buildings, and skyscrapers beginning in the mid-nineteenth century and drawing to a close, perhaps, with the collaboration of Skidmore Owings and Merrill architect Bruce Graham and engineer Fazlur Khan on the Sears Tower in 1974. The latter building remains one of the world's tallest, not because structural engineers cannot design a taller structure, but because there are very limited applications for a building more than 400 m high, and even fewer clients with the ability to finance its construction.

High technology is not as sexy as it was in the days preceding Three Mile Island and global warming, and what excitement remains is focused more on computers and biotechnology than on building technology. High tech's glamour has left the construction industry behind, and with it the desire of many architects to express the structural technology by which a building stands up. Visual expression of the structure is inherent to the design of tensioned fabric roofs, however. The beginning of the modern era in their construction might be dated to the completion of the first roof using permanent fiberglass fabrics. That occurred, ironically, in 1973, only a year before the completion of the Sears Tower and a waning of the interest in structural expressiveness among architects.

Conventional Organization of the Design and Construction Team

The architectural community's gingerly approach to fabric has created a void in the advancement of these forms. A few structural engineers have plunged into it to become not only the individuals advancing the technological state of the art but also primary creators of most of the outstanding individual designs. In order to do so, they have had to do their work in a manner at variance with firmly established construction industry practices.

There are many models for managing the design and construction of a building, and various ways of compensating each of the participants in order to provide incentives for creating the best product. In the majority of projects, the architect is the "prime" design professional responsible not only for determining the building's general configuration and design, but also for selecting and coordinating the work of the engineering consultants.

Typically, the structural engineer is the most critical of these consultants, and, in a well-managed project, the architect will work

The form of Eero Saarinen's TWA Terminal at Kennedy Airport derives more from sculptural considerations than structure. The inefficiency of form results in a thick and heavily reinforced shell.

closely with him to establish the general layout of the structural members, the allowances that must be made for mechanical and electrical systems, and other considerations.

The architect generally works with mechanical and electrical engineering consultants to establish the general requirements of the building's electrical, heating, ventilation, and air conditioning (HVAC) systems. They must keep the building interior within certain standards of temperature, humidity, and air flow. They must work reliably and efficiently, and they must fit within the restrictions imposed by the agreed-upon structural system and the limits of floor-to-floor height.

As the complexity of building technology expands, new fields of consulting continue to spring up. The team of specialists working with the architect of a major building project now typically includes consultants in structure, interiors, lighting, landscape, site work, signage, acoustics, window walls, roofing, elevators, mechanical, electrical, plumbing, fire safety systems, handicapped access, and cost.

The design of a building's structure is integrally linked with its wall locations, window sizes, room layouts, floor-to-floor heights, and economy. Because of this, the structural engineer has usually been first among equals on the architect's team of consultants. This remains so in spite of the increasing sophistication of mechanical and electrical systems and a growing sensitivity to energy use, because the design of the structural system has far greater impact on the building's form than any other element of its technology.

In the past, the architect who desired a building of unusual height, free span, or configuration had to work closely with the structural consultant from the beginning of the project, always seeking from him the limits of what was possible, pushing, cajoling, or pleading with the engineer to stretch the limits of his experience to achieve a result that best approximated the architect's vision of the building. Using advanced materials and analytical techniques, however, the contemporary structural engineer can make a structure of almost any size or configuration sufficiently strong and stable, although he may have to violate all rules of economy and simplicity of form in order to do so (see photograph on page 175). Ironically, improvement in the skill of structural engineers has decreased their influence on the design process, as the contemporary architect is less likely to ask his structural engineer "can you do it?" than "when will you have it done?"

While the architect configures a fascia built of glass and stone, aluminum, or concrete, his structural engineering consultant designs a steel or reinforced concrete supporting structure that will lie inside the fascia and support it (see photograph on page 177). As long as the structural engineer designs a system that is safe and economical, and as long as the structure stays within agreed-upon limits of dimension and configuration, the architect is seldom concerned with the details of the structural engineer's work. Mechanical, electrical, plumbing, acoustical, and other consultants work with similar independence, so long as they maintain the agreed performance, geometry, and other parameters.

The conventional form of the working relationship between the architect and the engineering consultants has advantages for the architect. He is, first of all, clearly established as the controlling member of the team, and because the roles and limits of responsibility of various members of the team are well defined, the architect's work in reviewing and coordinating the work of the consultants is minimized.

In this environment, the attitude of the architect and his consultants toward each other can sometimes be less than charitable. Engineers, for their part, often hold architects to be dreamy dilettantes whose ideas are not grounded in reality. Architects, in turn, may hold their engineers as a necessary evil, drones retained to address the technical issues inherent in realizing the architect's vision. They are seen not so much as collaborators as technicians who often must be browbeaten and pushed into making whatever accommodations are necessary to implement the architect's vision. The attitude is summed up by a sign in the restroom of a prominent architectural firm, a company that had recently won the American Institute of Architects' "Firm of the Year" award. I had been invited to the office to make a luncheon presentation to the architects about tensioned fabric structures. Visiting the men's room beforehand, I noticed a hand-lettered sign taped to the wall next to the urinals. In an architect's neat block letters, it said "ENGINEERS HERE," and the arrow below it pointed to the drain in the floor.

The Pittsburgh Plate Glass office tower is a monument to the potential of glazed fascia that offers no hint of the structure that supports it.

The role of the contractor and his relationship to the design team, to the owner, and to his subcontractors is similarly well defined in most contemporary construction. The contractor is retained by the owner, through a process of bidding or negotiation, to implement a design developed by the architect and his consultants. The contractor in turn retains a number of subcontractors to perform the work under his supervision. His work must typically pass specified tests and inspections in order to gain the acceptance of the design team, and any modifications he makes to the design typically require the review and approval of the architect or his consultants.

The precast concrete fascia of a typical modern office building is engineered and detailed by the precaster to conform to the geometric and performance parameters specified by the architect.

Organization of the Design and Construction Team for Tensioned Fabric Structures

There is a growing trend toward having contractors, through their subs, perform all or a portion of the work on a design/build basis, wherein the architect or one of his consultants provides general parameters and performance standards for a building component, and the subcontractor installing that component is responsible for its detailed design.

There are several reasons for this practice, currently common with such elements as fascia, fire sprinklers, and skylights. Most importantly, building technology has become so complex and diverse that few architects, even together with their consultants, have detailed knowledge of all the elements that compose a major building. Second, as architects and engineers are increasingly pressed by competition to minimize fees, they look to turn more of the detailed design work over to subcontractors and other parties as a means of maintaining profitability. Third, rising expectations for minimizing risk to property or building occupants are resulting in the need for steadily increasing documentation and validation of building design and construction. Much of this work has fallen on subcontractors as a matter of course.

The design/build approach is typically best applied to specialized building technologies in which skills and technical knowledge are not widespread, and in which prefabricated construction and proprietary technology are common (see the photograph on the previous page). It also finds important application in schedule-driven industrial applications where close coordination between designer and builder is particularly important.

Tensioned fabric roof structures represent a highly specialized building technology, one in which skills and technical knowledge rest in the hands of a small number of people, and in which prefabricated construction and proprietary technology dominate. And like other technologies of this type, they are often realized through some form of the design/build approach, in which the engineers and architects retained by the owner provide only schematic design and performance specifications, and the contractor must retain his own engineer to finalize a design that satisfies both the requirements of the owner's consultants and the particular capabilities and preferences of the contractor.

The vast majority of contemporary building construction is rectilinear in its geometry, and the general proportioning and configuration of structural members are fairly predictable to architects with even a modest knowledge of structural design. Since the building's structure is most often clad in some manner and hidden from view, the architect is typically not concerned with the appearance of structural elements or their detailing.

However, the means by which a fabric roof stands up, the manner in which it addresses energy use, light, and fire safety — are all inseparable from the way that it looks. Supporting masts typically are left exposed, and steel cables pass through space or lay against the fabric so that they remain visible from either above or below the roof. Even the layout of the seaming of the fabric is a strong visual element of the design. Natural light streams through the structure and artificial lighting and fire sprinkler systems are exposed below the ceilingless roof.

The high-tech movement of contemporary architecture provides a few examples of buildings with similarly exposed building systems, such as Paris' Pompidou Centre (see the photograph on page 181), the brilliant product of the collaboration among architects Renzo Piano and Richard Rogers and engineer Ove Arup. It is interesting to note that Pompidou and other outstanding high-tech buildings were not constructed according to a design/build model, but by unusually close collaborations between architects and distinguished consultants.

The unusual structural properties of the fabrics themselves have a great impact on the working relationship between the architect and the structural engineer. Due to their slenderness, fabrics typically have negligible resistance to either bending or compression. Because of these limitations in load-carrying ability, the fabric must be shaped in a very precise manner that allows it to carry all applied loads purely in tension. The determination of these shapes is both less commonplace and more complex than determination of the layout of a conventional concrete or steel frame, and the architect is typically dependent on a structural engineer specializing in tensioned fabric structures for assistance in determining the form of the roof.

Among the thousands of structural engineering firms practicing in the United States today, only a handful have staff members experienced in fabric structures, with knowledge of their specialized detailing, and versed in use of the non-linear finite element computer programs required to shape and analyze a tensioned fabric roof.

Some fabric structure contractors have worked to fill this void by providing their work on a design/build basis, either employing engineering staff or retaining one of the specialized consultants to offer services that range from initial consulting on design concepts and fabric capabilities to determination of precise shapes, stress analysis of fabric and supporting structures, and preparation of detailed cutting patterns for the fabric. This approach has several advantages. The architect and owner are assured of the engineering services of a party that presumably has extensive design experience with fabric and also is familiar with the unique construction requirements associated with the material. Retaining a single source for both design and construction, furthermore, offers at least the possibility for fewer conflicts over responsibility in the event of a problem in the finished product.

The design/build approach to fabric structures is not without its detractors, however, particularly in the United States, where there is a strong prejudice toward the quality control provided by engineers who work directly for the owner rather than the contractor to oversee and check construction work. Design/build work also becomes problematic for owners who wish to obtain multiple bids for a project rather than negotiate with a single contractor, as it is difficult for bidders to submit accurate quotations for work that has only been schematically designed.

It is a presumed loss of control over the final design that provides architects the greatest disincentive to acceptance of the

Exterior exposure of structural, mechanical, transport, and other systems at Pompidou Centre gives the look of a building turned expressively inside out.

Source: Enrique Limosner; used with permission.

design/build approach, however. The structural engineer on most building projects is well known to the architect, is employed by and paid by the architect, and subject to dismissal by the architect. A design/build contractor generally deprives the architect of all of these controls, however, and arouses the fear that the engineer may be more interested in designing the structure with an eye toward minimizing construction cost than maximizing construction quality.

Savvy contractors understand the prejudice many owners and architects have against design/build, and they work hard to overcome it. I asked an executive of Birdair, Inc., America's largest fabric structure contractor, about his company's design/build work, only to learn that use of the term itself is forbidden. "Wash your mouth out with soap," he chided. "We want to provide design only where it's required." His statement is literally true but ignores the fact that extensive design support is in fact required on nearly all of the structures that they build, and is more reflective of his firm's desire not to offend either architects or their consulting engineers than it is of Birdair's design capability and experience.

On other projects, the architect or owner retains a structural engineer with specialized knowledge of tensioned fabric structures. This engineer may design all structural elements for the project or may be retained only to structurally design the fabric membrane and related elements. In the latter case, the design of the foundations, walls, floors, and other conventional elements is left to a different structural engineer, typically one proximate to the building site and already familiar to the architect. The scope of the tension structure engineer's work may vary, too, from specification writing and schematic design of the roof, to full shaping and patterning of the fabric membrane and detailed design of supporting elements. In the former case, detailed design remains the responsibility of the fabric roof contractor, whose work is generally subject to the review of the consulting engineer.

Conclusions

Tensioned fabric roof structures may be designed either by the design/build approach or by an engineering consultant with specialized knowledge of this technology who is retained by the architect or the owner. The consultant may design the roof in its entirety or be retained only to provide general parameters and review of the roof contractor's detailed engineering. The appropriate formulation of the architect, structural engineer, and contractor team for completion of a given tensioned fabric roof ultimately will depend on the qualifications and experience of the parties involved as well as the characteristics of the project itself.

As noted earlier, the ability of contemporary structural engineers to achieve reliable results with structures that would previously have been considered experimental (see Chapter 5) is beginning to give architects a greater role in a medium dominated, until now, by engineers. As long as the number of structural engineers with experience in fabric was very small, and as long as the analytical tools used by them were limited in capability and not publicly available, architects were bound by the skills, intuitions, and preferences of the tension structure specialist with whom they chose to work. Tension structure design lived in the realm of magic, with a few engineering specialists the shamans able to bless or deny the architect's ideas about form. As long as the analytical tools available to these specialists were limited, design options were in turn limited, and fabric roof shapes tended to be restricted to those that were familiar and those having simple and readily predictable structural behavior. With contemporary tools, however, architects can ask their structural collaborators to evaluate

The tall masted canopy of Canada Place brings unabashed sail imagery to its waterfront site.

Source: Birdair, Inc.; used with permission.

unusual shapes or those having highly complex behavior, and can receive prompt and frequently favorable evaluations of roof forms that a pragmatically thinking engineer would never seek to use.

As advanced analytical tools and direct project experience have become more widespread, therefore, it seems likely that tensioned fabric roofs will take a step toward joining the mainstream of a construction industry in which architects act as form-givers and structural engineers are the professionals charged with addressing certain practical considerations of bringing those forms to reality. As architects begin to wrest some control of design back from the structural engineers, a different sensibility has begun to reflect itself in their designs. The maxims followed by structural engineers in tension structure design typically display a desire for simplicity of design, economy of material use, and redundancy that reflect the engineer's training and mind-set.

As architects have played a more critical role in the design of certain monumental tensioned fabric structures, their divergent motives have been expressed. In Vancouver's Canada Place (see the photograph on the previous page), the peaks of the masts have been lifted to an elevation far greater than that required to attain adequate fabric curvature, thereby significantly increasing the yardage of fabric required and the magnitude of lateral wind loads. The forms give a distinctly sail-like image that architect Eberhard H. Zeidler found appropriate to the waterfront structure. The enormous airport structure in Denver, Colorado (see the photographs on page 32), capped by a repeated module of double-masted tents, provides another relevant example. It was structural engineer Horst Berger's natural inclination to simplify design and maximize construction economy by making all 34 masts the same height. However, architect C.W. Fentress/J.H. Bradburn chose to provide masts of three different heights in an effort both to emphasize the junctions between the different terminals and to mirror the varying heights of the snow-capped Rocky Mountains visible in the distance.

The ideal model for formulating fabric roof designs may be a more richly collaborative process in which deference to the limitations posed by the divergent language, training, and conventional roles of architect and engineer are dropped. FTL/Happold, a merging of the New York architectural office of Future Tents Limited and engineering firm Buro Happold of Bath, England, has succeeded admirably in tensioned fabric roof design by working in this way. "At FTL/Happold, we have a 'barrier-free' design process," says principal Nicholas Goldsmith. "Our philosophy is 'let the best idea win.' It doesn't matter who said what" (*Fabric Architecture* 1999).

More than anything, though, the promoters of tensioned fabric roof construction must struggle uphill against historical images of tents as temporary, unreliable, dirty, and flammable construction. Against these obstacles, the tensioned fabric roof presents a variety of technological plusses that include durability, fire resistance, long-span capability, and natural daylighting. It is not, ultimately, technology that leads architects and owners to choose fabric, so much as the transcendent visual appeal of a medium in which expanses of material as smooth and white as porcelain cover space in curving forms of timeless grace.

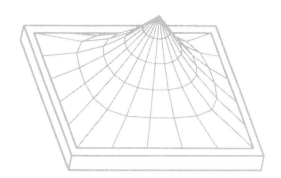

References

Architectural Record. 1984. "Tent Structures: Are They Architecture?" May.
Architecture. 1999. "Richard Rogers' Tensile Roofed Millennium Dome Brings London Under the Big Top." January, p. 109.
ASCE. 1996. "Air Supported Structures." *ASCE Standard 17-96*. Reston, Va., pp. 2–3.
———. 1997. "Structural Applications of Steel Cables for Buildings." *ASCE Standard 19-96*. Reston, Va., p. 4.
ASTM. 1975. "Standard Method of Test for Breaking Load and Elongation of Textile Fabrics." *D1682-59T*, West Conshohocken, Pa.
———. 1990. "Test Method for Failure in Sewn Seams of Woven Fabrics." *D1283-90a*, West Conshohocken, Pa.
———. 1996. "Standard Test Method for Resistance to Airflow as an Indication of Average Fiber Diameter of Wool Top, Card Silver, and Scoured Wool." *D1282-96*, West Conshohocken, Pa.
———. 1997. "Standard Test Methods for Coated and Laminated Fabrics for Architectural Use." *D-4851-97*, West Conshohocken, Pa.
———. 1999a. "Standard Test Method for Behavior of Materials in a Vertical Tube Furnace at 750°C." *E 136-99-e1*, West Conshohocken, Pa.
———. 1999b. "Standard Test Method for Cleanability of Surface Finishes." *C756-87*, West Conshohocken, Pa.
———. 1999c. "Standard Test Method for Coated Fabrics Abrasion Resistance (Rotary Platform, Double-Headed Abrader)." *D3389-94*, West Conshohocken, Pa.
———. 2000a. "Standard Test Methods for Fire Tests of Roof Coverings." *E 108-00*, West Conshohocken, Pa.

———. 2000b. "Standard Practice for Operating Enclosed Carbon Arc Light Apparatus for Exposure of Nonmetallic Materials." *G153-00ae1*, West Conshohocken, Pa.

———. 2001a. "Standard Test Method for Surface Burning Characteristics of Building Materials." *E 84-01*, West Conshohocken, Pa.

———. 2001b. "Standard Test Methods for Solar Energy Transmittance and Reflectance (Terrestrial) of Sheet Materials." *E424-71*, West Conshohocken, Pa.

———. 2002. "Standard Test Method for Effect of Household Chemicals on Clear and Pigmented Organic Finishes." *D1308-02*, West Conshohocken, Pa.

Basjanac, Vladimir. 1980. "Energy Analysis." *Progressive Architecture*. June, p. 121.

Beitin, Karl I. 1982. "Energy Performance of Fabric Roofs." *The Construction Specifier*. July, pp. 12-14.

Billington, David P. 1983. *The Tower and the Bridge*. New York: Basic Books.

Birdair. 1996. "Technical Specification and Fabric Characteristics." Amherst, N.Y.: Birdair, pp. 1-7.

BSI. 2000. "Testing Coated Fabrics." *British Standard 3424*, London: BSI.

Business Week. 1979. "A Fabric Roof Goes Retail." April 23.

Chemical Fabrics Corp. 1984. "Lindsay Park Aquatic Centre & Fieldhouse Project Profile." Merrimack, N.H.: CFC.

———. 1995. "Technical Data Sheets for Sheerfill Fabrics." Merrimack, N.H.: CFC.

———. 1996. "Technical Data Sheets for Sheerfill Fabrics." Merrimack, N.H.: CFC.

Crosby. 1998. Product literature. The Crosby Group, Inc.

Daugherty, H.B. 1999. "The Technology of Final Tensioning Fabric Structures." *Structural Engineering in the 21st Century: Proceedings of the 1999 Structures Congress*. Reston, Va.: ASCE.

DePaola, Edward. 1994. Untitled paper presented at the ASCE 1994 Structures Congress held in Atlanta, Georgia.

Dery, Marcel. 1992. "Improved Composite Materials for Architectural Structural End Use." Presented at the 1992 Industrial Fabrics & Equipment Exposition, Phoenix, Ariz.

DIN. 2000. "Testing of Plastic Films—Tear Test Using Trapezoidal Test Specimen with Incision." *DIN53363*, Berlin, Germany, DIN.

———. 2002. "Testing of Artificial Leather; Tensile Test." *DIN53354*, Berlin, Germany, DIN.

DuPont. 1983. "Performance Specification Characteristics: Longevity of Tedlar-Clad Vinyl-Coated Polyester Fabric for Tension Fabric Structures." DuPont Company.

Engineering News Record. 1998. "Setting Record for 2000." July 20, pp. 26–31.

Fabric Architecture. 1999. "Cross-continental (Design) Currents." March/April, pp. 54–59.

Fabric Structures International. 1997. "Manufacturer's Data for Softglass Fabrics." Norcross, Ga.

Ferrari. 1999. Fabric Structures Series 8000 product literature. Cedex, France.

FTMS. 1978. "Strength of Cloth, Tearing; Tongue Method." *191/5134*, Washington, D.C.: General Services Administration.

———. 1989. "Strength and Elongation, Breaking of Woven Cloth; Grab Method." *191/5100*, Washington, D.C.: General Services Administration.

Geiger, David. 1989. *Encyclopedia of Architecture: Design, Engineering, & Construction, Volume 3*. New York: John Wiley & Sons, pp. 418–422.

Gorman, John. 1992. "A Hot Topic." *Fabrics & Architecture*. May/June, pp. 44–45.

Grossman, Susan. 1991. "The Topcoating Option." *Fabrics & Architecture*. January/February, pp. 37, 38.

Harvie, Gregor. 1995. "The Science of Shelter." *Fabrics & Architecture*. July/August, pp. 16–23.

Hatton, E.M. 1979. *The Tent Book*. Boston: Houghton Mifflin, p. 53.

Holgate, Alan. 1997. *The Art of Structural Engineering*. Stuttgart/London: Axel Menges, p. 68.

Huntington, Craig G. 1984. "Visual Expression in Fabric Tension Structures—The Possibilities and Limitations." *Proceedings of the 1984 International Symposium on Architectural Fabric Structures*. Architectural Fabric Structures Institute, pp. 137–141.

———. 1986. Personal review of University of La Verne Campus Center energy use records.

———. 1987. "Permanent Architectural Fabric Structures—Performance of the New Materials Technology." *Construction and Building Materials*. Surrey, England: Scientific and Technical Press Ltd.

———. 1989. "Manipulation of Shell and Fabric Roof Form." In *Steel Structures*, Jerome S.B. Iffland, ed. New York: ASCE, pp. 706–715.

———. 1990. "Fabric Structure Developmental Needs." Structures Congress Abstracts of the 1990 Structures Congress New York: ASCE, pp. 29–30.

———. 1992. "Tensioned Fabric Structures." *Construction India, 1992 Annual, Indian & Eastern Engineer*, pp. 87–89.

———. 1993. "Methodologies and Technologies for Tensioned Structures." *l'Arca*. Milan, Italy: *l'Arca* Edizioni. July/August, pp. 82-83.

———. 1994. "Fabric Roof Design Responds to New Technologies." *Spatial, Lattice and Tension Structures: Proceedings of the IASS-ASCE International Symposium 1994.* New York: ASCE, pp. 674–683.

———. 1995. "Within Limits." *Fabrics & Architecture.* January/February, pp. 17–21.

———. 1997. "Fabric Structure Pretensioning Mechanisms." *Fabrics & Architecture.* May/June, pp. 35–38.

Ishii, Kazuo. 1995. *Membrane Structures in Japan.* Tokyo: SPS Publishing Company, pp. 163, 181, 369.

Ito, M. 1991. "Cable Stayed Bridges: Recent Developments and Their Future." New York: Elsevier Science, p. 58.

Japanese Standards Association. 1999a. "Testing Methods for Woven Fabrics." *JIS L1096*, Tokyo, Japan: JSA.

———. 1999b. "Rubber Coated Fabrics." *JIS K6328*, Tokyo, Japan: JSA.

Kadolph, Sara J., and Langford, Anna L. 1998. *Textiles*, 8th Ed. Upper Saddle River, N.J.: Prentice Hall, p. 219.

Lambert, Kenneth. 1986. "Performance Requirements for Architectural Membrane Fabrics." *Proceedings of LSA '86.* Lightweight Structures Association of Australia, p. 841.

Liddell, Ian. 1989. "The Engineering of Surface Stressed Structures." *Patterns 5.* Bath, England: Buro Happold Consulting Engineers, p. 9.

Lodewijk, T. 1967. *The Way Things Work, Volume 1.* New York: Simon & Schuster.

Moulder, Ron, and Merrill, Jim. 1982. "Acoustical Properties of Glass Fiber Roof Fabrics." Presented at the 104th Meeting of the Acoustical Society of America. Melville, NY. Acoustical Society of America.

Murrell, V. William. 1984. "Tensile and Tear Strength Criteria for Architectural Fabric Structures." Roseville, Minn.: Industrial Fabrics Association International, pp. 76–81.

National Fire Protection Association. 1999. "Standard Methods of Fire Tests for Flame Propagation of Textiles and Films." *Standard 701.* Quincy, Mass.: NFPA.

Nervi, Pier Luigi. 1955. *Structures*, Giuseppina and Mario Salvadori, trans. New York: F.W. Dodge Corp.

Oberdick, Willard A., Boonyatikarn, Soontorn, and Wu Her-Fu. 1981. "Energy Performance of Fabric Roof Structures." Ann Arbor, Mich.: Architectural Research Laboratory, College of Architecture & Urban Planning, pp. 1–28.

Owens-Corning. 1979. Internal company correspondence, October 23.

Rebeck, Gene. 1991. "Sound Structures." *Fabrics & Architecture.* March/April, pp. 42–43.

San Jose Mercury. 1981. "Bullock's Raises the Roof in San Mateo." September 14, p. 5D.

Schwitter, Craig. 1994. "The Use of ETFE Foils in Lightweight Roof Construction." *Spatial, Lattice and Tension Structures: Proceedings of the IASS-ASCE International Symposium 1994.* New York: ASCE.

Seaman Corporation. 1984. "The Architectural Fabrics of the Future." Millersburg, Ohio.

Shaeffer, R.E. 1996. *Tensioned Fabric Structures: A Practical Introduction.* New York: ASCE, pp. 1-4, 8-1–8-5.

Sidenius, Derek. 1982. "Dome Slashed, Principal Irate." *Capital Times-Colonist.* July 6, p. 9.

Sinofsky, Mark. 1985. "Thermal Performance of Fabric in Permanent Construction," pp. 5–7. Presented at the Seminar on New Developments in Fenestration Systems, ASHRAE Annual Conference. Atlanta, Ga.: ASHRAE.

Textile World. 1970. "Textile Fibers Comparison Table." August, p. 9.

Uniform Building Code. 2000. International Conference of Building Officials. 1997. Whittier, Calif.: ICBO.

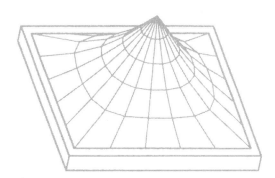

Project Credits

Avery Aquatic Center
Stanford, California
Architect: ELS/Elbasani & Logan Architects
Structural Engineer: Rutherford & Chekene
Fabric Canopy Contractor: Sullivan & Brampton
Fabric Canopy Engineer: Huntington Design Associates, Inc.
Completed: 2001

Baypointe Station
San Jose, California
Architect: SBA Architects
Structural Engineer: Lee Incorporated
Fabric Canopy Contractor: Birdair, Inc.
Fabric Canopy Engineer: Huntington Design Associates, Inc.
Completed: 2000

Buchser School
Santa Clara, California
Architect: Steinberg Group
Structural Engineer: Huntington Design Associates, Inc.
Roof Contractor: unbuilt

Buckingham Palace Ticket Office
London, England
Architect: Sir Michael Hopkins & Partners
Structural Engineer: Ove Arup and Partners
Roof Canopy Contractor: Landrell Fabric Engineering, Ltd.
Completed: 1994

Buena Ventura Shopping Center
Ventura, California
Architect: Charles Kober Associates
Fabric Roof Contractor: Birdair, Inc.
Fabric Roof Consultant: Huntington Design Associates, Inc.
Completed: 1984

Bullock's of Northern California
San Mateo, California
Architect: Gene Zellmer
Structural Engineer: Geiger Berger Associates
Fabric Roof Contractor: Birdair, Inc.
Completed: 1981

Burnside Residence
Tesuque, New Mexico
Architect: Archaeo Architects
Structural Engineer: Sonalysts, Inc.
Fabric Canopy Engineer: Huntington Design Associates, Inc.
Fabric Canopy Contractor: Rader Awning & Upholstering, Inc.
Completed: 2003

Canada Place
Vancouver, British Columbia
Architect: Zeidler Roberts
Structural Engineer: Geiger Berger Associates
Fabric Roof Contractor: Birdair, Inc.
Completed: 1986

Carlos Moseley Pavilion
New York, New York
Architect: FTL Associates
Structural Engineer: M.G. McLaren, P.C.
Fabric Roof Contractor: Sky Structures
Membrane Engineer: Buro Happold
Completed: 1989

Carleton Centre
Ottawa, Ontario
Canopy Designer: Duvall Design
Completed: 1997

Chang Sha Amphitheater
Chang Sha, China
Architect: Beijing N & L Fabric Technology
Structural Engineer: Huntington Design Associates, Inc.
Fabric Roof Contractor: Beijing N & L Fabric Technology
Completed: 1998

Cincinnati Bell Telephone
Cincinnati, Ohio
Architect: FRCH Worldwide
Membrane Designer & Contractor: Duvall Design
Completed: 1998

Denver Airport Terminal
Denver, Colorado
Architect: C.W. Fentress, J.W. Bradburn & Associates
Structural Engineer: Horst Berger (with Serverud Associates)
Fabric Roof Contractor: Birdair, Inc.
Completed: 1994

Electronic Arts
Redwood City, California
Architect: Skidmore, Owings & Merrill
Structural Engineer: Nishkian & Associates
Roof Contractor: P.J.'s Canvas
Membrane Fabricator: Fabric Structures International
Fabric Roof Engineer: Huntington Design Associates, Inc.
Completed: 1998

El Grande Bigo
Genoa, Italy
Architect: Renzo Piano
Structural Engineer: Ove Arup
Fabric Roof Contractor: Canobbio
Completed: 1992

El Monte Resort
Taos, New Mexico
Architect: Dharma Properties
Executive Architect: Nims, Calvani & Associates
Structural Engineer: MacCornack Engineering
Roof Contractor: Eide Industries
Fabric Roof Engineer: Huntington Design Associates, Inc.
Completeed: 2003

Florida Festival
Orlando, Florida
Architect: Robert Lamb Hart Architects
Structural Engineer: Geiger Berger Associates
Roof Contractor: Birdair, Inc.
Completed: 1980

Folkestone Chunnel Terminal
London, England
Architect: Building Design Partnership
Fabric Roof Contractor: Birdair, Inc.
Completed: 1992

Georgia Dome
Atlanta, Georgia
Architect: Heery Architects & Engineers, Inc.
Structural Engineer: Weidlinger
Fabric Roof Contractor: Birdair, Inc.
Completed: 1991

Good Shepherd Church
Fresno, California
Architect: Gene Zellmer
Fabric Roof Contractor: Birdair, Inc.
Completed: 1982

Hajj Terminal
Jeddah, Saudi Arabia
Architect and Structural Engineer: Skidmore, Owings & Merrill
Roof Contractor: Owens-Corning Fiberglass w/ Birdair, Inc.
Fabric Roof Engineer: Geiger Berger Associates
Completed: 1981

Hanover Park Recreation Facility
Hanover Park, Illinois
Architect: The Shaver Partnership
Structural Engineer: Bob D. Campbell
Fabric Roof Contractor: Owens-Corning Fiberglass Corporation
Completed: 1976

Harrah's Carnaval Court
Las Vegas, Nevada
Architect: Henry Conversano Associates
Structural Engineer: Huntington Design Associates, Inc.
Fabric Roof Contractor: Sullivan & Brampton
Completed: 1997

Hollywood & Highland
Hollywood, California
Architect: The Ehrenkrantz Group
Executive Architect: Altoon & Porter
Structural Engineer: Robert Englekirk
Fabric Canopy Contractor: Sullivan & Brampton
Fabric Canopy Engineer: Huntington Design Associates, Inc.
Completed: 2001

Independence Mall Pavilion
Philadelphia, Pennsylvania
Architect: H2L2
Structural Engineer: Geiger Berger Associates
Fabric Roof Contractor: Birdair, Inc.
Completed: 1976

Jameirah Beach Hotel
Palm Court, Sundeck Stair, Sundeck Bar, Coffee Collonade, and End Canopies
Dubai, United Arab Empirates
Architect: W.S. Atkins
Fabric Roof Contractor: Birdair, Inc.
Fabric Canopy Engineer: Huntington Design Associates, Inc.
Completed: 1996

Kaleidoscope Shopping Center
Mission Viejo, California
Architect: Altoon & Porter
Structural Engineer: ANF & Associates
Canopy Contractor: Advanced Structures, Inc.
Membrane Fabricator: Eide Industries
Fabric Canopy Engineer: Huntington Design Associates, Inc.
Completed: 1998

King Fahd Stadium
Riyadh, Saudi Arabia
Architect: Ian Fraser
Structural Engineer: Schlaich Bergerman & Partners
Design Engineer: Horst Berger Partners
Fabric Roof Contractor: Birdair, Inc.
Completed: 1986

Lindsay Park Sports Centre
Calgary, Alberta
Architect: Chandler Kennedy Architectural Group
Structural Engineer: Geiger Berger Associates
Fabric Roof Contractor: Birdair, Inc.
Completed: 1984

Marin Technology Center
San Rafael, California
Architect: Robinson Mills & Williams
Structural Engineer: Steven Tipping & Associates
Fabric Canopy Contractor: P.J. Canvas
Fabric Canopy Engineer: Huntington Design Associates, Inc.
Completed: 1996

Milano Fair Ground
Milan, Italy
Architect: Luciano Sgalbazzi
Structural Engineer: Massimo Majowicki
Fabric Roof Contractor: Canobbio
Completed: 1986

Millennium Dome
London, England
Architect: Richard Rogers Partnership
Structural Engineer: Buro Happold
Fabric Roof Contractor: Birdair, Inc.
Completed: 1999

Montreal Olympic Stadium
Montreal, Quebec
Architect: Roger Taillibert
Structural Engineer: Regis Trudeau & Associates, ABBDL; Trudeau, Gascon, Lalancette & Associates
Fabric Roof Contractor: Birdair, Inc. (for replacement of original roof)
Completed: 1987

Munich Olympic Park Swimming Pool
Munich, Germany
Architect: Günter Behnisch & Partners with Frei Otto
Structural Engineer: Leonhardt & Andra
Fabric Roof Contractor: Krupp, Waagner-Biro, Voest
Completed: 1972

Munich Olympic Stadium
Munich, Germany
Architects: Günter Behnish & Partners with Frei Otto
Structural Engineer: Leonhardt & Andra
Contractor: Krupp, Waagner-Biro, Voest
Completed: 1972

Munich Ice Rink
Munich, Germany
Architect: Ackermann and Partner
Structural Engineer: Schlaich Bergerman & Partners
Roof Contractor: Mauer Söhne, Munchen Pfeifer, Memmingen Koch, Rinsting
Completed: 1985

National Semiconductor Amphitheater
Santa Clara, California
Architect: Bellagio Associates
Structural Engineer: Huntington Design Associates, Inc.
Fabric Roof Contractor: Sullivan & Brampton
Completed: 1995

National Semiconductor Entry Canopy
Santa Clara, California
Architect: Bellagio Associates
Structural Engineer: Huntington Design Associates, Inc.
Fabric Canopy Contractor: Sullivan & Brampton
Completed: 1994

Park City Mall
Lancaster, Pennsylvania
Architect: Cope Linder Associates
Structural Engineer: Cagley & Harman
Fabric Roof Contractor: Birdair, Inc.
Completed: 1989

Parkwest Medical Office Building
Knoxville, Tennessee
Architect: King & Johnson
Fabric Canopy Contractor: Eide Industries
Fabric Canopy Designer & Engineer: Huntington Design Associates, Inc.
Completed: 2001

Pavilions at Buckland Hills
Manchester, Connecticut
Architect: Cambridge Seven Architects
Structural Engineer: Weidlinger Associates
Fabric Roof Contractor: Birdair, Inc.
Completed: 1989

Pizzitola Athletic Facility
Brown University
Providence, Rhode Island
Architect: The Eggers Group PC
Structural Engineer: Geiger KKBNA
Fabric Roof Contractor: Birdair, Inc.
Completed: 1989

Pontiac Silverdome
Pontiac, Michigan
Architect: O'Dell/Hewlett & Luckenbach, Inc.
Structural Engineer: Geiger Berger Associates
Fabric Roof Contractor: Birdair, Inc.
Completed: 1975

Prophet's Mosque
Medina, Saudi Arabia
Architect: Architektürburo Dr. Bodo Rasch
Structural Engineer: SL Sonderkonstruktionen und Leichtbau GmbH
Fabric Canopy Contractor: SL Sonderkonstruktionen und Leichtbau GmbH
Fabric Canopy Engineer: Buro Happold
Completed: 1992

Radome
Various locations
Structural Engineer: Walter Bird
Fabric Roof Contractor: Birdair, Inc.

Realmonte Sporting Club
Rome, Italy
Fabric Roof Contractor: Canobbio
Completed: 1994

Roman Amphitheater Cover
Nimes, France
Architect: Finn Geipel & Nicolas Michelin
Structural Engineer: Schlaich Bergerman & Partner
Fabric Roof Contractor: Stromeyer, Konstanz CENOTEK, Greven
Completed: 1988

San Diego Convention Center
San Diego, California
Architect: Arthur Erickson
Structural Engineer: Horst Berger Partners
Fabric Roof Contractor: Birdair, Inc.
Completed: 1989

Sea World Whale Pool
Orlando, Florida
Architect: Sea World
Structural Engineer: Advanced Structures, Inc.
Fabric Canopy Contractor: Fabritec Structures
Completed: 2000

Sherway Gardens Mall
Etobicok, Manitoba
Architect: Zeidler Roberts
Structural Engineer: Dowdell, Pall, Ellis & Associates
Fabric Roof Contractor: Birdair, Inc.
Completed: 1989

Tampa Airport Hanger No. 1
Tampa, Florida
Architect: Rowe, Holmes Associates
Fabric Roof Contractor: Birdair, Inc.
Completed: 1981

Tokyo Dome
Tokyo, Japan
Architect/Structural Engineer: Nikken Sekkei & Takenaka Corporation
Fabric Roof Contractor: Takenaka Corporation
Completed: 1988

Tropicana Dome
St. Petersburg, Florida
Architect: Hellmuth, Obata & Kassabaum
Structural Engineer: Geiger Engineers
Fabric Roof Contractor: Birdair, Inc.
Completed: 1989

United States Pavilion
Osaka, Japan
Architect: David Brody
Structural Engineer: Geiger Berger Associates
Fabric Roof Contractor: Taiyo Kogyo Corporation
Completed: 1970

University of La Verne Campus Center
LaVerne, California
Architect: The Shaver Partnership
Structural Engineer: Bob D. Cambell, Lin Kulka Yang & Associates
Fabric Roof Contractor: Birdair, Inc.
Completed: 1973

Walden Galleria
Buffalo, New York
Architect: Dal Pos Associates
Fabric Roof Contractor: Birdair, Inc.
Completed: 1988

Weber Point Events Center
Stockton, California
Fabric Canopy Designer & Engineer: Huntington Design Associates, Inc.
Fabric Canopy Contractor: Sullivan & Brampton
Completed: 1999

West German Pavilion
Montreal, Quebec
Architect: Frei Otto, Rolf Gutbrod
Structural Engineer: Leonhardt & Andra
Fabric Roof Contractor: L. Stromeyer & Company
Completed: 1967

White Lotus Restaurant
Hollywood, California
Architect: Umemoto Associates
Fabric Canopy Designer & Engineer: Huntington Design Associates, Inc.
Fabric Canopy Contractor: Eide Industries
Completed: 2002

Wilgenberg Greenhouse
Reedley, California
Structural Engineer: Huntington Design Associates, Inc.
Fabric Roof Contractor: Flexys Structures, Inc.
Completed: 1993

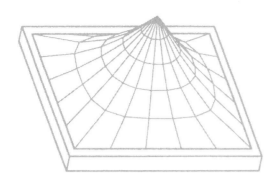

Index

abrasion test 61, 64
Ackermann & Partner 41, 194, Plate 16
acoustics 166–169
acrylic 71
acrylic panels 54
adhesion test, coating 64
Advanced Structures, Inc. 193, 195
aerodynamics 44
air-inflated lenses 48–49
air-inflated structures 47, 96
air pressure, internal 44
air-supported structures 22, 29, 43–47, 96, 137
Altoon & Porter 193, Plate 1, Plate 3
American Society for Testing and Materials (ASTM) 60, 62
American Society of Civil Engineers 63, 93, 114
amphitheaters 9, 24, 25, 29, 34, 48, 49, 107, 111, 115, 117, 125, 129, 130, 138–140, 147, 149, 150, 169, 170, 192, 194–196, Plate 12, Plate 13
analytical procedures 87–89
ANF & Associates 193
anticlastic surface 14, 96–97
arch stability 40
arch-supported shape 15
arch systems 38–42; erection 147–148
Archaeo Architects 191, Plate 2
arches, canted 38–39; crossed 38; steel pipe 38
architect, role of 173–184

Architekturburo Dr. Bodo Rasch 195
Arup, Ove 180, 192, Plate 15
aspect ratio, excessive 19
assembly structures 21
asymmetry 136
Atkins, W.S. 34, 193, Plate 10
Avery Aquatic Center 36, 117, 191, Plate 23

Baypointe Station 149, 191, Plate 7
bearing plates, mast-base connections 125
Behnish, Gunter 54
Beijing N&L Fabric Technology 34, 192
Bellagio Associates 25, 194, Plate 12
Bellusch, Anthony 7
bending elements 83–84
Berger, Horst 32, 34, 35, 77, 184, 192, Plate 14
biaxial tensile test 61, 63
Billington, David 8
Bird, Walter 43, 47, 98, 195
Birdair, Inc. 43, 46, 73, 90, 96, 98, 119, 141, 150, 163, 181, 191, 192, 193, 194, 195, 196, Plate 6, Plate 7, Plate 8, Plate 9, Plate 10, Plate 14, Plate 18, Plate 19, Plate 20, Plate 21, Plate 22, Plate 24
Bradburn, Fentress 32
Brody, David 195
Buchser School 128, 129, 191
Buckingham Palace Ticket Office 84, 191

Buena Ventura Shopping Center 39, 135, 191
building codes 94–95, 170. *See also* codes and standards
Building Design Partnership 192, Plate 6
Bullock's Department Store 157, 165, 171, 191
Burnside Residence 191, Plate 2
Buro Happold 31, 184, 192, 194, 195, Plate 19

C.W. Fentress, J.W. Bradburn & Associates 184, 192, Plate 9
cable curvature, inadequate 16–17
cable fittings 82–83; closed end 109–11; jaw end 109–110; stud end 109–110, 112; speltered 109–110, 111; swaged 109–110, 111, 112
cable-suspended masts 33
cable-tensioning 126–128
cable-to-cable connections 115–116
cable trusses, radiating 51
cables 82–83, 94; anchorages 108; catenary 16, 104–105, 107, 108, 111–112, 148–149; circumferential hoop 50; diagonal 50; dome 22, 29, 50–52, 97; edge 107; fabrication of 145–146; length 145; loop 120; net 53–55, 86, 91; saddle 113–116; span-to-sag ratio 16; steel 146; tension 17, 115
Cagley & Harman 194
Cambridge Seven Architects 194
Campbell, Bob D. 193
Canada Place 183, 184, 192
Canobbio 192, 194, 195, Plate 15, Plate 17
cantilever canopies 35–37
Carleton Centre 6, 192
Carlos Moseley Pavilion 169, 170, 192
cathedrals 3, 5
Chandler Kennedy Architectural Group 163, 193
Chang Sha Amphitheater 34, 129, 138–140, 192
Charles Kober Associates 191
Chemfab. *See* St. Gabain Performance Plastics
Chemical Fabrics Corp. *See* St. Gabain Performance Plastics
Cincinnati Bell Telephone 9, 192
clamp bars 106

clamp plates 103
cleaning 151, 152, 153–154
coating materials 68–69, 78
codes and standards 60, 62, 63, 64, 66, 67, 72, 78, 93–94, 114, 170–171
cold weather 162
compensation properties, fabric 134
compression members 11, 83–84; long 19
computerized techniques 89–92
condensation 163–164
cones 14, 22–27; applications, range of, 24; drawbacks 23; inverted 16, 27, 37; mast-supported 22; simple 16
conical roofs, erection 147
connections 101–103; cable-to-cable 115–116; mast-top 116–121; mast base 122–125
construction phases 151–153
construction team, conventional 174–176; tensioned fabric structures 178–182
contemporary building construction 178–179
contractor, role of 173, 177
Cope Linder Associates 194
corners, terminations 105–106
cost considerations 67–68; films 79; polyester 72; polytetra-fluoroethylene (PTFE) 76; silicone coatings 68–69, 78
crimp interchange 59–60
cuff material 104
cuffs 140, 144
curvature 7, 13, 93; inadequate 15–17; requirements 21; reversals 19
cutting 143–145

Dal Pos Associates 196
Davies, Michael 31
daylighting 155–158, 161
deflation 46–47; prevention of 47
deformations 92
Denver Airport Terminal 32–34, 105, 158, 184, 192, Plate 9
design, art of 15–20; principals of 12–14
design/build approach 178–179
design evaluation 92–94
design methodologies 85
design safeguards 46
design team, conventional 174–175; tensioned fabric structures 178–182

Dharma Properties 192, Plate 11
di Reno, Casalecchio Plate 17
dimensional instability, polyester 70
dirt accumulation 72, 151, 153–154, 159–160
domed roofs, conventional 45
Dowdell, Pall, Ellis & Associates 195
Duball Design 192
Dupont 73
durability 64–66; cables 83; films 79; polyester 70–71; polytetrafluoroethylene (PTFE) 74–76; silicone coatings 77
Duvall Design 192
Duvall, Charles 6

East Area Health Center 164
edge clamping, adjustable 126, 127
Effenberger, John 58
Eggers Group PC 194
Ehrenkrantz Group 193, Plate
Eide Industries 192, 193, 194, 196, Plate 1, Plate 4, Plate 5, Plate 11
Eiffel, Gustave 6
El Grande Bigo 7, 115, 192, Plate 15
El Monte Resort 105, 106, 119, 192
Electronic Arts building 8, 192
ELS Architects 36
ELS/Elbasani & Logan Architects 36, 191, Plate 23
energy efficiency 160–165; measuring 164–165
Englekirk, Robert 193
Erdman, Lee 96
erection 147–151
Erickson, Arthur 34, 195, Plate 14
ETFE. See ethylene-tetra-flouroethylene polymer
ethylene-tetra-flouroethylene polymer (ETFE) 78–79
European Neutral (EN) Standards 60, 62
eye terminations 109, 111

fabric coatings 68, 78
fabric creep 70, 154
fabric curvature, inadequate 15–16
fabric, direct tensioning of 126
fabric durability 64–66, 70–71, 74–77, 79
fabric joints 103–104
fabric materials, characteristics 58
fabric mechanical properties 60–63; films 78–79; polytetrafluoroethylene (PTFE) coated 73–74; silicone coated 77

fabric membrane, effective life span 65–66
fabric panels 51–52
fabric performance parameters 59–68
fabric restraint 106, 108
fabric shipment 137–138
fabric stiffness behavior 60, 133
fabric tensioning 125–132, 147–151
fabric terminations 103–108, 117, 140
fabric weave 58–59
fabric width 134–137
Fabric Structures International 192
fabrication 143–146
FabriTec Structures 195, Plate 1
Ferrari 70
fiber damage 62
fiberglass fabrics 73–78; polytetrafluoroethylene (PTFE) coated 45, 57, 64, 73–76, 80, 108, 119, 135, 145, 156; silicone coated 76–78
fiberglass substrate 19
films 78–79
fire resistance 66–67, 72, 169–172; films 79; polytetrafluoroethylene (PTFE) coated 76; PVC-coated polyester 72; silicone coated 78
fire testing 67
flex fold tests 62; silicone coatings 77
Flexys Structures, Inc. 196
Florida Festival 120, 192
flying masts 34
Folkestone Chunnel Terminal 22, 192, Plate 6
force components 18
Ford, Jim 150
form, variety of 21
forms, defining 98–99
foundations 152
Fraser, Ian 193, Plate 22
FRCH Worldwide 192
freeze-thaw damage 65
FTL Associates. See FTL Happold
FTL/Happold 184, 192
Fuller, Buckminster 99
Future Tents Limited. See FTL Happold

Geiger Berger Associates 30, 40, 53, 90, 163, 191, 192, 193, 195, Plate 8, Plate 18, Plate 20
Geiger cable domes 50, 51–53
Geiger, David 44, 45, 50, 99

Geiger KKBNA 194
Geipel, Finn 195
Georgia Dome 51–52, 192, Plate 21
German Pavilion, Montreal 42
Good Shepherd Church 136, 193, Plate 24
Gore, W.L. 80
grab tensile test 61, 62–63
Graham, Bruce 174
Gunter Behnisch & Partners 194
Gutbrod, Rolf 196

H2L2 193
Hajj Terminal, Jeddah International Airport 1, 2, 29–30, 31, 90, 193, Plate 8
Hanover Park Recreation Facility 193
Harrah's Carnival Court 23, 24, 193
Heery Architects & Engineers, Inc. 192, Plate 21
Hellmuth, Obata & Kassabaum 195, Plate 20
Henry Conversano Associates 193
high-strength steel wire rope 82
Hollywood & Highland Ballroom Canopy 115–116, 124, 159, 193, Plate 3
Horst Berger Partners 193, 195, Plate 9, Plate 22
Hubbell, Kent 164
humidity 164
Huntington Design Associates, Inc. 191, 192, 193, 194, 196, Plate 1, Plate 2, Plate 3, Plate 4, Plate 5, Plate 7, Plate 10, Plate 11, Plate 12, Plate 13, Plate 23
hypalon 68
hypar-tensegrity roofs 51–53
hyperbolic paraboloids 96

Independence Mall Pavilion 121, 122, 193
inspection 151
installation plan 146–147
Institute for Lightweight Structures, University of Stuttgart 42

Jameirah Beach Hotel 34, 35, 108, 124, 132, 193, Plate 10
Japanese Standards Association (JSA) 60
John Hancock Building, Chicago, Illinois 6

Kaleidoscope Shopping Center 27, 193, Plate 1

Kennedy, Chandler 40
Kevlar 79, 81
Khan, Fazlur 6, 174
king post design 22, 23, 50
King & Johnson 194
King Fahd Stadium 35–36, 120, 121, 122, 193, Plate 22
knife damage 65
knits 81
Kober, Charles 39
Krupp, Waagner-Biro, Voest 194

L. Stromeyer & Company 196
Landrell Fabric Engineering, Ltd. 191
layout 146–147
Lee Incorporated 191
Leonhardt & Andra 54, 194, 196
Levy, Matthys 51–53
lighting 158–159
light transmission 71–72, 155–157, 161–163; fabrics 66; films 79; polytetrafluoroethylene (PTFE) coated 76; silicone coated 78
lightning hazard 166
Lindsay Park Sports Centre 39–4, 163, 193
liner membranes 163, 164
load transfer 119
loading 94–96
loads, dead 94; earthquake 95; heavy 94; live 94–95; point 94; snow 47, 95, 166; vertical 87; wind 13, 95

M.G. McLaren, P.C. 192
MacCornack Engineering 192
Maillart, Robert 6, 8
maintenance 153–154
Majowiecki, Massimo 28, 194, Plate 17
Marin Technology Center 126, 193
mast bases 122–125, 128–130; single-degree-of-freedom 123, 124
mast jacking systems 128–131
mast tensioning 128–132
mast termination, inadequate 17–19
mast-top connections 116–122, 150–151
mathematical models 141–142
Mauer Sohne, Koit-Werk Plate 16
Mauer Sohne, Munchen Pfeifer, Memmingen Koch, Rinsting 194
Membrane Structures Association of Japan 62, 63, 65

membranes, deformability 94; non-structural 127
mesh 81, 82
Michelin, Nicolas 195
Milano Fair Ground 28, 194, Plate 17
Millennium Dome 29, 30, 31–32, 194
Montreal Olympic Stadium 122, 123, 194
Munich Ice Rink 28, 41–42, 54, 55, 194, Plate 16
Munich Olympic Park Swimming Pool 87, 89, 194
Munich Olympic Stadium 54, 89–90, 194

National Gallery of Art, Washington, D.C. 3
National Semiconductor Amphitheater 24, 25, 107, 115, 117, 130, 194, Plate 12
National Semiconductor Entrance Canopy 26–27, 131, 194
neoprene 68
Nervi, Pier Luigi 3, 6, 8
Nikken Sekkei & Takenaka Corporation 195
Nims, Calvani, & Associates 192, Plate 11
Nishkian & Associates 192
Nohmura, Moto 47, 58

O'Dell/Hewlett & Luckenbach, Inc. 195, Plate 18
Otto, Frei 53, 54, 86, 118, 194, 196
Ove Arup and Partners 191
Owens-Corning Fiberglass Corporation 30, 73, 75, 167, 168, 193, Plate 8

P.J.'s Canvas 192
Palm Court Canopies 34
Park City Mall 135, 194
Parkwest Medical Office Building 37, 108, 194, Plate 5
patterning 141–143
Pavilions at Buckland Hills 97, 194
Pei, I.M. 3
Piano, Renzo 179, 192, Plate 15
Pittsburgh Plate Glass office tower 177
Pizzitola Athletic Facility 38, 194
Place Stadium, Vancouver, B.C. 47
polyester film 78–79
polyester, mechanical properties 70; polyvinyl chloride (PVC) coated 64, 68–72, 74, 134, 138, 144, 156
polyester webbing 82
polyvinyl chloride (PVC) foil, heat shrunk 86
polyvinylidene fluoride (PVFD) 70–71
Pompidou Centre, Paris 179–180, 181
ponding 95, 166
Pontiac Silverdome 46, 169, 195, Plate 18
prestress shape 96; changing 93
pretensioning mechanisms 125–132
Prophet's Mosque, Medina, Saudi Arabia 80, 195
PVC. See polyvinyl chloride
PVFD. See polyvinylidene fluoride

Rader Awning & Upholstering, Inc. 191, Plate 2
Radome 43, 47, 195
Realmonte Sporting Club 43, 195
Regis Trudeau & Associates, ABDL 194
repair kits 153
reverberation characteristics 166–169
Richard Rogers Partnership 31, 194, Plate 19
ridge and valley systems 32–34, 40, 148
Robert Lamb Hart Architects 192
Robinson Mills & Williams 193
Roebling, John 6
Rogers, Richard 180
Roman amphitheater cover 48, 49, 195
Rowe, Holmes Associates 195
Rutherford & Chekene 191

Saarinen, Eero 175
saddles 14; shapes 96; catenary boundary 15; curving boundary 15; linear boundary 15
safety factors 92–93, 112, 148–149
sag 17, 89
San Diego Convention Center 33, 34, 195, Plate 14
SBA Architects 191, Plate 7
scalloped forms 32
Schlaich Bergerman & Partners 193, 194, 195, Plate 16, Plate 22
Schlaich, Jörg 3, 41, 49, 54–55, 103
Sea World (architect) 195

Sea World Whale Pool Canopy 113, 195
seam direction 135, 136
seam selection 133–140
seam tensile strength tests 63
Seaman Corporation 70
seaming 143–145
seams 103–104; cemented 144; continuous 135; heat-sealing 144–145; orientation of 135–137; sewn 144
Sears Tower 174
semi-cylindrical shapes 43
Serverud Associates 192, Plate 9
Sgalbazzi, Luciano 194
shaping 86–87, 96–98
shaping technicques, non-numerical 86–87
Shaver Partnership 193
shell structures 2
Sherway Gardens Mall 134, 195
shopping malls 1, 27, 39, 135
silicone coating 76
Silverdome, Pontiac, Michigan 45–46
Sir Michael Hopkins & Partners 191
Skidmore, Owings & Merrill 174, 192, 193, Plate 8
Sky Structures 192
slope, effect on acoustics 169
snow-melting systems 166
soap film models 86, 91
soiling 65, 66
solar properties 160, 162
Sonalysts, Inc. 191
Sonderkonstruktionen and Leichtbau GmbH 195
sound absorption 167
sound transmission 166, 168
Spencer Secondary School 65
sprinkler systems 171–172
St. Etienne Cathedral 5
St. Gabain Performance Plastics 58, 73
St. Mary's Cathedral, San Francisco, California 3
stability 7, 19
stabilizing cables 121
stadiums 21, 45–47, 50
stainless steels 112–113
standards. *See* codes and standards
steel, bending radius 114
steel structural strand 82
steel supporting elements 83–84
steel tensile rods 82
Steinerg Group 191
Stephen C. O'Connell Center 165

Steven Tipping & Associates 193
storms 46
stress/strain properties 63; fabrics 60
stressing 96–98
strip tensile strength 92; polytetrafluoroethylene (PTFE) 73
strip tensile test 60, 62, 61
Stromeyer, Konstanz CENOTEK, Greven 195
structural analysis software 92
structural art 8
structural engineer, role of 174–176, 181
Stuttgart Garden Fair roof 3
suction forces 13
Sullivan & Brampton 111, 191, 193, 194, 196, Plate 3, Plate 12, Plate 13, Plate 23
support conditions 93
supporting elements 83–84, 145–146
supporting structure, unstable 19
suspended roofs 28–32
suspension cables 121
synclastic surface 13, 14

Tailliber, Roger 194
Taiyo Kogyo Corporation 58, 195
Takenaka Corporation 47, 195
Tampa Airport Hanger No.1 111, 118, 130, 195
tear strength 60
tears 50, 55
temperature variations 95
tennis bubbles 43, 45
tennis facility, Hanover Park, Illinois 38
tensegrity domes 51–53
tensile forces, cables 109, 113
tensile strength, polyester 70
tensile strength tests 60–63
tension forces, radial 116–117
tension measuring devices 150
tension structures 11
tensioning 148–151
tensioning rods 117
tent shapes 86
tents, circus 4, 5, 169; military 5; nomadic 5; traditional 1, 57, 69, 97
terminations, cable 108–116; cable, choice of 110; cable, designing 110–113; at catenaries 104–105; supporting structures 105
thermal properties 161, 162

three-dimensional truss systems 50
tie-back cable-tensioning 127
timber 84
Tokyo Dome 47, 195
tongue tear test 61, 63–64
topcoats 70–71
towers 122
translucency. *See* light transmission, fabrics,
trapezoid tear test 61, 63
Tropicana Dome, Florida 51, 195, Plate 20
Trudeau, Gascon, Lalancette & Associates 194
turnbuckles 106, 107, 110, 112, 124, 151
TWA Terminal, Kennedy Airport 175

U-strapping 104, 105
ultraviolet radiation exposure 65, 75
ultraviolet radiation protection 69, 71
Umemoto Associates 196
uniaxial loads 60
United States Pavilion, Osaka 31, 44, 195
University of La Verne Student Center 65, 68, 73, 74, 75, 119, 153, 162, 163, 165, 172
urethane 68–69, 71

vandalism 65–66
ventilation 118–119; heated 164

Walden Galleria, Buffalo, New York 5, 196
warp fibers 58, 136
weather 165–166
webbing, polyester 108
Weber Point Events Center 9, 25, 26, 111, 125, 147, 148, 150, 196, Plate 13
Weidlinger Associates 51, 99, 192, 194, Plate 21
welded seams 63
West German Pavilion 87, 196
White Lotus Restaurant 35, 196, Plate 4
Wilgenberg Greenhouse 196
wind forces 44, 95
wire mesh 82
wrinkling 149–150

Zeidler, Eberhard H. 184
Zeidler Roberts 192, 195
Zellmer, Gene 191, 193, Plate 24